MAPPING HAZARDS IN NEPAL'S MELAMCHI RIVER
CATCHMENT TO ENHANCE
KATHMANDU'S WATER SECURITY

NOVEMBER 2023

ASIAN DEVELOPMENT BANK

CONTENTS

TABLES, FIGURES, AND MAP

Map

Forewords

Nepal has a long history of disasters caused by earthquakes, floods, and landslides. The 2021 dual flood events, triggered by the impacts of climate change and exacerbated by vulnerable Himalayan geology, led to massive landslides and debris floods in the Melamchi River catchment area. Approximately 17 people were killed and downstream settlements and infrastructure devastated, including the headworks of the Melamchi Water Supply Project (MWSP) that was ready to be commissioned in September 2021.

The MWSP plays a central role in ensuring the drinking water security, health, and well-being of the people of the Kathmandu Valley. Damage to the long-awaited project highlighted the vulnerability of Nepal's infrastructure and underscored the efforts of the Government of Nepal to reduce disaster risk and protect its development gains.

It is vital to ensure that all necessary measures are taken to insulate important projects like the MWSP against future catastrophic hazards, particularly as climate change takes effect.

This technical report is an important step in understanding the causes and effects of disasters like those seen in the Melamchi River catchment. It identifies the root causes of the Melamchi debris flows and lays down the groundwork for efforts to repurpose or reconstruct the critical infrastructure destroyed by the 2021 events.

Apart from supporting the government's efforts to quickly restore the MWSP, this report will serve as a useful guide that outlines the measures that need to be considered to ensure the resilience of key infrastructure both in Nepal and neighboring countries. Its findings and recommendations will resonate with a wide range of policymakers, practitioners, engineers, and researchers on the need to design resilient infrastructure projects that help drive sustainable development.

I would like to thank the Asian Development Bank (ADB) for supporting and promoting this important report and my colleagues at the Ministry of Water Supply (MWS) and the Melamchi Water Supply Development Board

(MWSDB) for their help and guidance. I hope it will motivate Nepal's government departments, agencies, districts, and local municipalities to fully consider the impacts of hazards in all their activities and strategize suitable measures to best mitigate them.

I believe this report will help decide the best steps to take to both protect the lives and livelihoods of people living in the Melamchi catchment and ensure the long-term delivery of essential water supplies to Kathmandu.

Ram Adhar Sah
Secretary
Ministry of Water Supply
Government of Nepal

Forewords

For the fast-growing population of the Kathmandu Valley, the project that brings drinking water from the Melamchi River to their homes, public buildings, and businesses, is a vital lifeline that plays a crucial role in equitable and sustainable development. As such, it is important to prioritize the resilience of the Melamchi Water Supply Project (MWSP) and protect it against the impacts of myriad hazards such as earthquakes, floods, debris flows, and landslides.

The 2021 flood events destroyed MWSP's critical headworks structures that were built to divert water to Kathmandu, burying them under 15-20 meters of debris. This not only highlighted risks to the capital's water security as climate change threatens to complicate conditions, the floods also emphasized the urgent need to understand threats from the river's catchment and develop mitigation measures.

With this objective, the Asian Development Bank (ADB) has worked closely with the Government of Nepal, the Ministry of Water Supply (MWS), and the Melamchi Water Supply Development Board (MWSDB) to carry out this first part of a comprehensive two-stage study. This initial report identifies the multiple hazards present in the Melamchi River catchment area and analyzes factors that transformed them into cascading disasters during the 2021 monsoon.

The report recommends mitigation measures to stabilize the river catchment and reduce the vulnerability of key infrastructure along the river's course. However, it shows that focusing on infrastructure alone is not enough and underlines the need to work with a wide range of stakeholders to institutionalize evidence-based multi-hazard risk assessments from the project planning stage. It also recommends the rollout of comprehensive early warning systems to help affordably mitigate the impacts of climate-induced disasters in the highly fragile Himalayan region.

The Melamchi incident illustrated that "building back better" cannot remain a mere catchphrase. The 2021 floods exceeded the design flood of the original Melamchi headworks, and new design considerations must ensure that the repurposed or rehabilitated headworks can cope with similar as well as greater events.

Working in remote parts of the Himalayan region is challenging, and I would like to extend my thanks to the Government of Nepal, my ADB colleagues, and the technical assistance team for having undertaken the extensive research that formed the basis for this study.

I hope this comprehensive and practical report will guide and assist all stakeholders in operationalizing the MWSP, both to guarantee Kathmandu's water security and show how to incorporate resilience into future Himalayan projects.

Arnaud Cauchois
Country Director
Nepal Resident Mission
Asian Development Bank

Acknowledgments

This technical report was initiated by the Government of Nepal, the Ministry of Water Supply (MWS), and the Melamchi Water Supply Development Board (MWSDB), with the technical assistance of the Asian Development Bank (ADB) in the wake of the 2021 Melamchi River basin floods.

Production of the report was led by Saugata Dasgupta, senior project management specialist, ADB; and Sujan Raj Regmi, project analyst, ADB. It was prepared under the guidance of Ram Adhar Sah, secretary, MWS; Arnaud Cauchois, country director, ADB Nepal Resident Mission.

The report benefited from the support and input of specialists at Nepal´s National Disaster Risk Reduction Management Authority, the Department of Hydrology and Meteorology, and the Department of Mines and Geology. Valuable support and guidance were also provided by the International Centre for Integrated Mountain Development, and a specialist World Bank team.

ADB's technical assessment team was led by Hans Enggrob, river morphology expert. Team members included Sanjiv Shah, deputy team leader and structural engineer; Eckhard Schnäcker, engineering geologist; Raghunath Jha, geographic information system expert; Ranjan Kumar Dahal, geological expert; Gyanendra Lal Shrestha, geotechnical expert; Bishnu Raj Baral, sediment transport expert; Vishnu Pandey, hydrology expert; Mahesh Bhattarai, water supply expert; and Shalini Patwal, ADB consultant.

The report was peer reviewed by BGC Engineering, an international consulting firm.

ABBREVIATIONS

ADB	Asian Development Bank
CCTV	closed-circuit television
DEM	digital elevation model
DGSD	deep-seated gravitational slope deformation
DHM	Department of Hydrology and Meteorology
DMG	Department of Mines and Geology
DRM	disaster risk management
EOP	Emergency operation procedure
ERT	Electrical resistivity tomography
EWS	early warning systems
GCP	ground control point
GIS	geographic Information System
GLOF	glacial lake outburst flood
HEC-RAS	Hydrologic Engineering Center - River Analysis System
ICIMOD	International Centre for Integrated Mountain Development
IRBM	integrated river basin management
km	kilometer
kN/m^2	kilonewton per square meter
LDOF	landslide dam outburst flood
LEWS	landslide early warning system
LIDAR	Light Detection and Ranging
masl	meters above sea level
m^3	cubic meter
m^3/s	cubic meter per second
mm/hr	millimeters per hour
MRWDS	Melamchi River Water Diversion Subproject
MWS	Ministry of Water Supply
MWSDB	Melamchi Water Supply Development Board
MWSP	Melamchi Water Supply Project
NDRRMA	National Disaster Risk Reduction and Management Authority
O&M	operation and maintenance
PMOF	probable maximum outburst flood

PMP   probable maximum precipitation
PSD   particle size distribution
SAR   Synthetic Aperture Radar
SRT   seismic refraction tomography
SWAT   Soil and Water Assessment Tool
TA   technical assistance
WMO   World Meteorological Organization
WTP   water treatment plant

EXECUTIVE SUMMARY

Nepal's Ministry of Water Supply (MWS) launched the Melamchi Water Supply Project (MWSP) almost 2 decades ago to ease chronic water shortages in Kathmandu. Supported by the Asian Development Bank (ADB), it was designed to divert 170 million liters of raw water a day from the Melamchi River through a 26-kilometer (km) water tunnel, where it would be treated and supplied to the capital. However, before full project operations could begin in September 2021, two catastrophic debris floods damaged the headworks of the Melamchi River Water Diversion Subproject (MRWDS). This rendered the system that controls the extraction of water from the river unusable.

Triggering Factors

The debris floods resulted from a series of cascading events in and around the Melamchi River catchment area. In 2015, the 7.8-magnitude Gorkha earthquake triggered numerous landslides in the catchment and increased its susceptibility to slope instabilities. In June 2021, following rainfall and a sharp temperature rise, a glacial lake outburst flood[1] (GLOF) in the upper river catchment caused massive erosion of a major sediment deposit at Bhemathan, 16 km upstream of the headworks.

Combined with precipitation, the debris floods generated a major landslide at Melamchi Ghyang which blocked the river. The buildup of water behind this natural dam caused a landslide dam outburst flood[2] (LDOF) which sparked the formation and subsequent collapse of a series of new natural dams. As a result, vast quantities of debris were eroded from the upstream river channel and deposited at the headworks and further downstream. The floods also aggravated an existing landslide at the headworks.

Impacts of Debris Floods

Extreme flooding in 2021 caused a number of deaths; displaced hundreds of people; and washed away homes, roads, and bridges. At the headworks and in the upstream catchment, the events caused the following:

- Triggering of new landslides and reactivation of existing landslides in the catchment.
- Significant bank erosion and riverbed scour in the upper river stretches, major sediment deposition at the headworks and further downstream.
- Burial of the entire headworks area under 15-20 meters of debris.
- Extensive damage to the headworks structures and equipment.
- Aggravation of an existing landslide at the headworks.

---

1   Caused by the failure of a natural dam of ice, soil, or rock containing meltwater from a glacier.

2   Triggered by the failure of a dam that was formed by a landslide.

As a consequence, the fragile catchment became more vulnerable to future landslides and erosion, and the river is now exposed to further natural damming and outburst floods. The headworks landslide poses a serious threat to the structural safety of the headworks, and a major runout of the landslide could block the river and cause outburst flooding.

Map 1: Location of the Melamchi River Basin

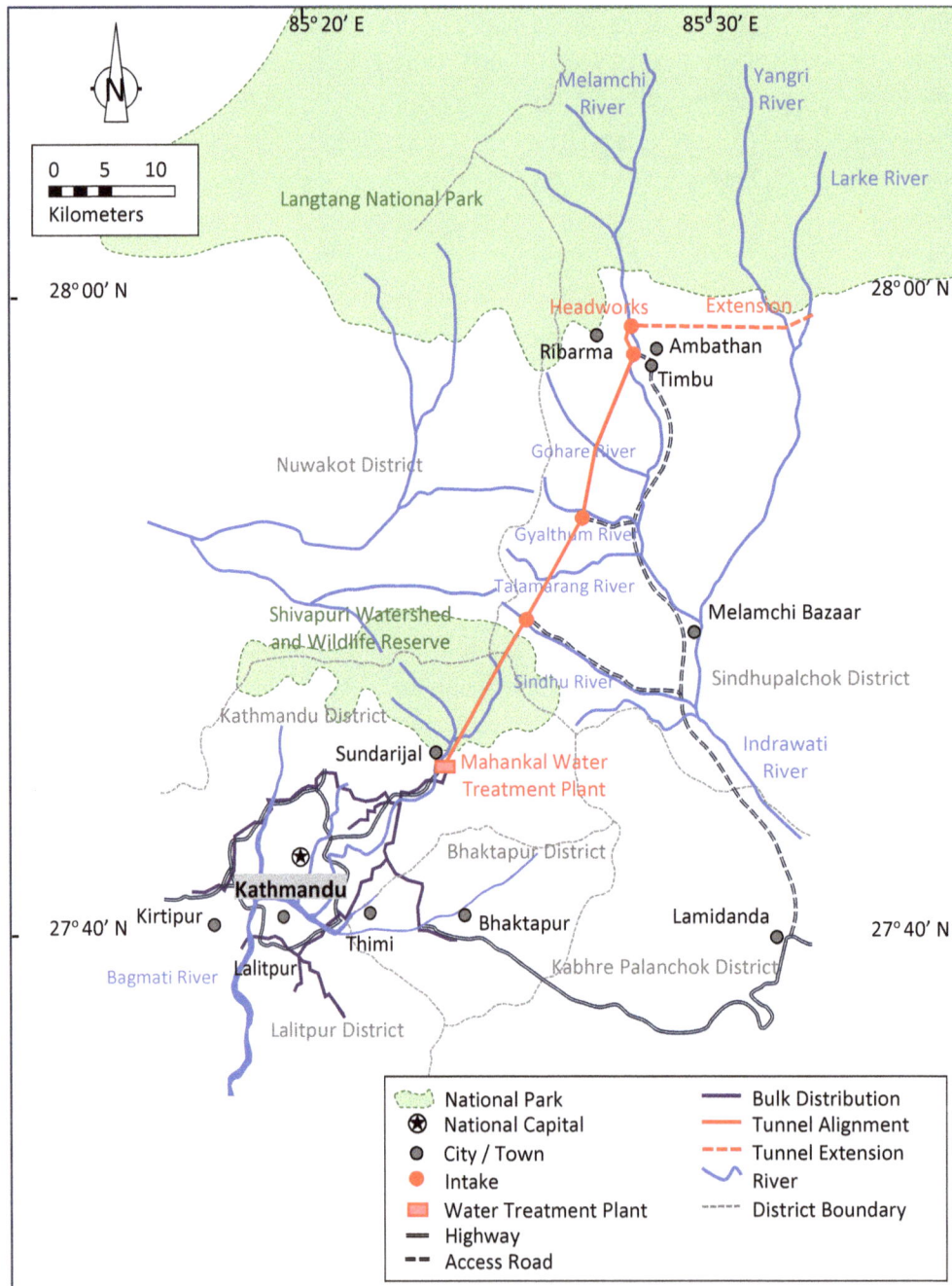

Source: Authors and Melamchi Water Supply Development Board.

The Studies

The 2021 events underscored the MWSP's vulnerability to extreme events. In view of this, ADB initiated a detailed technical assessment to identify the causative factors and impacts of the floods and develop measures to ensure the resilience of Kathmandu's drinking water supplies. The major findings of the assessment were the following:

- The flood events were associated not only with rainfall runoff but multiple hazards that combined and interacted with each other to domino into cascading effects.
- The floods had an estimated peak discharge of 2,700 cubic meter per second (m³/s), with flood waves that travelled at 22 meters per second (m/s) and reached the headworks in 12 minutes.
- Ten glacial lakes and at least two glaciers are present in the Melamchi catchment, but only a few lakes pose an outburst threat.
- Fifteen landslides in the catchment are critical, of which five, including ones at the headworks, Bhemathan, and Melamchi Gyang, are highly vulnerable.
- The area from Bhemathan to the headworks had a very high multi-hazard value from earthquakes, GLOF, LDOF, rainfall, heat waves, snowmelt, landslides and landslide damming, debris and mudflows, erosion, and accretion.
- The geomorphological processes that led to the 2021 events, including landslides, sediment deposition and erosion, etc., will continue to occur in the catchment over the foreseeable future.

Impacts of Climate Change

Using global climate models, the studies also projected the climate of the Melamchi catchment for the mid-future (2051 to 2075) and far future (2076 to 2100) with reference to the baseline (1992 to 2021). These models showed that the average annual maximum temperature of the catchment would increase by 2.6°C in the mid-future, and by 4.3°C in the far future. The corresponding average annual precipitation would rise by 13% and 29% for the respective periods.

Models also showed the average annual river discharge would increase by 7% in the near future (2021 to 2050), 19% in the mid-future, and 36% in the far future, with minor increases in the dry season (March to May) and substantial in the monsoon season (June to September).

Furthermore, the studies indicated that temperature rise would accelerate the retreat of snow, ice, and glaciers, and contribute to increased volumes of glacial lakes that may subsequently generate GLOFs. A warmer climate would also raise the permafrost[3] line and cause water stored as ground ice in the permafrost region to thaw, exacerbating risks of hazards.

---

3   Defined as soil or rocky ground, including ice or organic material, that remains at subzero temperature for at least 2 consecutive years.

New Criteria and Approach for Maximum Flood Estimation

When the Melamchi headworks plans were drawn up around 20 years ago, they were designed for a 10,000-year return period flood of magnitude 600 m^3/s. This flood frequency was selected considering the importance of the headworks, and the flood discharge was estimated using conventional methods practiced in the early 2000s when the project was designed.

However, the design flood was exceeded by as much as 4.5-fold in the 2021 events, mainly because conventional estimation methods were based purely on rainfall-runoff data. The methods also referred to a period when climate change was a non-issue.

Exceeding the design flood highlighted the need for incorporating climate change impacts in estimation of maximum floods by including various processes in geology, geomorphology, hydrology, etc., that influence multi-hazards. Accordingly, the concept of probable maximum outburst flood (PMOF) was introduced in this study. This concept involves estimations of GLOF, LDOF, cloudburst[4] floods, and probable maximum flood (PMF) for the catchment and the selection of the peak flood discharge based on a rational combination and comparison of the various floods.

Application of the PMOF approach to the Melamchi catchment yielded a maximum flood value of 3,500 m^3/s. Given the various possible scenarios of multi-hazard combinations in the catchment, the uncertainty in the PMOF estimate thus derived can be very high.

Mitigation Measures

The present study proposes the following mitigation measures to reduce the hazard intensity of extreme debris and flash floods in the Melamchi catchment:

- Future headworks designs consider a peak flood discharge and sediment concentration that incorporates hazards and climate change impacts.

- Comprehensive monitoring of the river and catchment and establishment of a multi-hazard early warning system (EWS).

- Setting up of a disaster risk management (DRM) plan that also involves local residents in river monitoring.

- Introduction of solutions to reduce landslide risks and offer broad community cost-benefits.

- Establishment of an integrated river basin management plan to identify, prioritize, and implement local hazard protection solutions at vulnerable sites.

Countries of the Himalayan region face challenges similar to those in the Melamchi catchment. Therefore, to better understand the climate change impacts at a regional level, it is strongly recommended that multi-hazard assessment and planning tools, such as climate risk assessment

---

4   An extreme amount of precipitation which occurs over a small area in a short period of time when moist air travels up a mountainous region.

guidelines and development control norms, be institutionalized, and knowledge repositories be developed for the benefit of all. These measures will help develop economical, coherent, and integrated strategies to deal with complex climate change issues.

Looking Ahead

For the MWSP to fulfill its objectives, comprehensive steps are needed to ensure that hazards do not threaten future water supplies. A second phase of this study will, therefore, develop feasible solutions that enhance resilience in the headworks, the river and its catchment. It will recommend climate-resilient designs to rebuild the headworks and propose intermediate supply solutions. Its recommendations will be critical for reducing disaster risk for residents and to future-proof the capital's water security.

This study highlights the importance of regional collaboration and river basin-level initiatives for improving climate and disaster risk management and paving the way for a climate-resilient future. Nepal has been encouraged to continue with integrated river basin management (IRBMs) for critical river basins, including the Melamchi basin. ADB is also planning to provide technical assistance on interventions at a regional level (including the Hindu Kush Himalaya Climate Risk Management Project) based on the outcome of this study. Guidelines that will emerge from this initiative will provide a well-founded platform for developing new projects in the region.

1. INTRODUCTION

Project Background

The Government of Nepal initiated the Melamchi Water Supply Project (MWSP) in 2001 to alleviate the chronic shortage of potable water in the fast-urbanizing Kathmandu Valley. Implemented by the Melamchi Water Supply Development Board (MWSDB), the project was designed to transfer 170 million liters of raw water daily from the Melamchi River to a water treatment plant (WTP) on the outskirts of the valley, from where treated water would be distributed to residents through a new transmission system and an upgraded distribution network.

The water transfer component of the MWSP was developed under the Melamchi River Water Diversion Subproject (MRWDS) using the following core structures:
- Headworks located 35 kilometers (km) northeast of Kathmandu and consisting of an ungated weir with an intake to divert the river water and a gravel trap and desanding basin to remove sediment from the diverted water.
- An alternative diversion system, formed by a coffer dam, a diversion tunnel and ductile iron pipes, for diverting the river water during maintenance of the headworks.
- A 26 km long tunnel for conveying the diverted water from the headworks to the WTP.

By June 2021, the project's headworks were in an advanced stage of completion, with major parts of their structures constructed and their hydromechanical equipment and instrumentation installed. The initial filling of the water tunnel was successfully completed on 27 March 2021, and the MWSDB commissioned the WTP as well as the transmission and distribution systems in the valley with water conveyed through the tunnel.

On 15 June 2021, the MWSDB started the dewatering of the tunnel for its inspection. It aimed to complete the remaining works at the headworks and the water tunnel, including any rectifications identified during its inspection, and commission the MRWDS in its entirety by September 2021.

Events and Their Impacts

Just before their scheduled commissioning, the MRWDS headworks were severely impacted by two major flood and debris flow events in the Melamchi River. The first event occurred on 15 June 2021 after heavy rainfall and glacial lake outburst flooding (GLOF) in the upper reaches of the river catchment. This was followed by a similar event on 1 August 2021 that was caused by rainfall and a landslide dam outburst flood (LDOF) in the Melamchi catchment.

The flooding and debris flow events were results of various cascading disaster events in and around the Melamchi catchment. The origin of these events can be traced back to the

7.8-magnitude 2015 Gorkha Earthquake that increased the geological fragility of the catchment and its susceptibility to land instabilities. The subsequent precipitation and sharp temperature rise during the 2021 monsoon generated a GLOF in the upper reaches of the catchment that triggered the collapse of a large pre existing sediment deposit in the river channel at Bhemathan, about 16 km upstream of the headworks.

Toe erosion caused by the flows from Bhemathan, coupled with prolonged rainfall in the catchment, triggered a major landslide further downstream that dammed the river. Breaching of this natural dam generated outburst floods from the lake that formed behind it, and this, in turn, created new natural dams whose breach caused additional outburst floods. Cumulatively, the outbursts eroded the riverbanks between Bhemathan and the headworks area, and deposited sediments in the headworks area and further downstream. Toe erosion due to these floods and increased soil saturation due to prolonged rainfall also aggravated an existing landslide on the left bank of the headworks.

Figure 1: Events Leading Up to the 2021 Melamchi Basin Floods

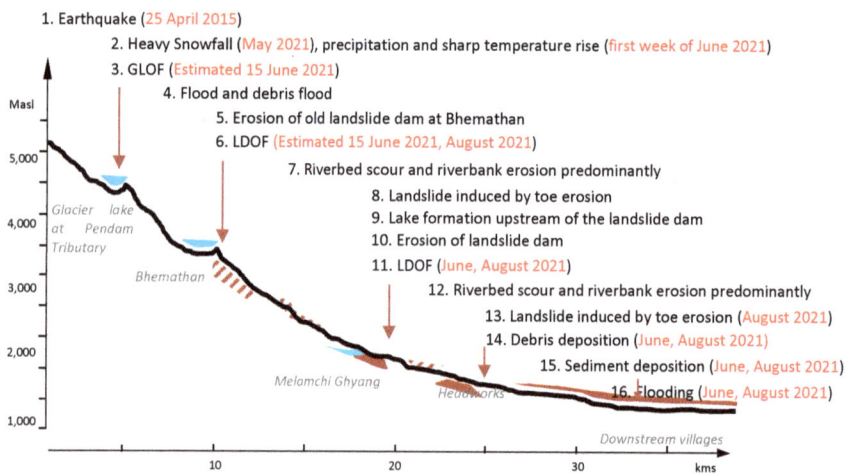

GLOF = glacier lake outburst flood, LDOF = landslide dam outburst flood, kms = kilometers, masl = meters above sea level.

Source: Authors.

The event of 15 June 2021 substantially changed the local river morphology, depositing sand, gravel, and boulders in some river stretches and eroding the riverbed and banks in others. It inundated the headworks area and buried the headworks structures, including the construction diversion tunnel and an adit[5] further downstream, under a 15–20 m deposit of sand, gravel, boulders, and debris. It also washed away the alternative diversion system coffer dam, permanent bridges at the headworks and the adit, as well as the contractors' camps and facilities, causing considerable loss of life and property.

---

5   A short horizontal or sub-horizontal tunnel driven from the surface to the main tunnel to provide access during its construction and operation.

Pre-disaster (above) and post-disaster condition (below) of Melamchi headworks (photo by TA team)

The second event also caused heavy erosion along the river channel, with extensive collapse of the V-shaped channel walls. In addition, it severely damaged roads, bridges, and public and private property in the downstream reaches of the river.

Aerial surveys of the Melamchi catchment following the events showed the downstream parts of the large sediment deposit at Bhemathan being eroding rapidly under successive river run offs and rainstorm events. Inspection of parts of the headworks structures that were gradually exposed by the river flows showed the structures, and their hydromechanical equipment and instrumentation, had suffered extensive damage from the abrading, rolling, and pounding actions of boulders. Structures, or parts thereof, that remained buried under the debris were also expected to have undergone varying degrees of damage due to the action of the boulders. Fortunately, the water tunnel was not affected as its inlet gate had been closed for dewatering and inspection just a few hours before the first event.

The debris at the headworks was partially removed prior to the 2022 monsoon, but parts of the headworks structures remained buried. The cleared areas were refilled with debris by 2022 floods, albeit to a lesser extent.

Like most Nepali river catchments, the Melamchi catchment is steep[6] and characterized by the presence of glacial lakes, fragile geology, and natural and human-induced disturbances such as settlements and construction. This makes the catchment susceptible to landslides, natural damming, dam breaches, and subsequent sediment morphological issues including deposition and erosion[7]. Hence, there is an elevated probability that such events may recur, and the risks may be exacerbated due to the large amount of mobile sediment present in the river which will eventually be transported further downstream.

View of the Melamchi River gorge with headworks in lower right corner (photo by TA team).

Objectives and Scope of Assessment Study

The 2021 debris floods in the Melamchi River highlighted the vulnerability of the MRWDS to extreme climate-change-induced events and underlined the need to identify and implement measures to enhance its resilience and sustainability. A first step toward this effort was an assessment of the damage to the headworks structures to ascertain the possibility and means

---

6   The bed slope varies from 1:6 to 1:13, or 8% to 17%, between the Bhemathan knickpoint which is at an elevation of 3,500 m, and the headworks 16 km downstream at an elevation of 1,430 m.

7   International Center for Integrated Mountain Development (ICIMOD). 2021. *The Melamchi Flood Disaster: Cascading Hazard and the Need for Multi-hazard Risk Management*. Kathmandu: ICIMOD.

of rehabilitating them so that they could continue functioning according to their design intent. In consideration of this, ADB engaged a multi-disciplinary team of experts to provide technical assistance (TA) and carry out the following studies and assessments on the Melamchi catchment and the headworks structures:

1. Identify the factors that triggered the 2021 Melamchi floods and debris flows.
2. Conduct detailed assessment of the impacts of the flood and debris flow events on the river morphology, riverbed, valley slope stability, and headworks structures.
3. Study the feasibility of long-term measures for remediation and stabilization of the river morphology. Recommend appropriate and cost-effective solutions.
4. Identify potential land instability and erosion hazards in the Melamchi catchment affecting the safe operation of the headworks. Study the feasibility of managing them and recommend appropriate, and cost-effective solutions.
5. Conduct detailed damage assessment of the MRWDS headworks and develop short-, medium-, and long-term measures for their proper hydraulic functioning and adequate structural integrity.
6. Undertake climate change analysis based on different models and assumptions and identify its impact on flow patterns, floods, droughts, erosion, and sedimentation.
7. Identify alternative intake design options or upgrades that may be required to adapt it to the upstream catchment degradation and improve its resilience.
8. Propose comprehensive strategic options for the medium– to long–term integrated recovery and adaptation of the MRWDS, focusing on the intake and upstream catchment with ranking based on cost, resilience, and technical complexity.

The studies and assessments were planned in two phases[8] as indicated in Table 1.

Table 1: Scope of Consulting Services, Phasing, and Key Tasks

Phase 1	Catchment and River Studies
A	River morphological studies
B	Hazard mapping, catchment land, and slope stabilization.
C	Identification of suitable measures for treating the upstream river catchment
Phase 2	Damage Assessment, Rehabilitation, and Upgrading of MRWDS
A	Damage assessment, including assessment of the possibility of rehabilitation and repurposing of the headworks structures and installations
B	Improvement of hydraulic performance and sediment handling at the headworks
C	Identification of strategic options for medium- and long-term integrated recovery/adaptation of MRWDS, ranked on cost, resilience, and technical complexity

MRWDS = Melamchi River Water Diversion Subproject.
Source: Authors.

8   Asian Development Bank and Melamchi Water Supply Development Board. 2022. *Damage Assessment of Melamchi Headworks* and *Hazard Mapping of the Catchment, TA6596-NEP, Inception Report.*

The tasks to be performed in Phase 2 were slightly modified following the results of Phase 1. The modifications included the following:

1. Additional surveys and investigations of the landslide near the headworks to provide a basis for the conceptual design of stabilization solutions.
2. Development of conceptual redesign of the headworks.

Summary of Methodology

The Phase 1 assessment had three study areas:

- The **catchment**, which is the source of water and sediments in the river channel.
- The **river channel**, which transports water and sediments to the headworks area.
- The **headworks structures,** which divert the river water, exclude sediments from it, and convey the sediment-free design discharge to the water tunnel.

The methodology for the study included:

Field Surveys
- Topographical surveys
- Geological mapping
- Geophysical surveys
- Sediment sampling and laboratory testing

Catchment Studies
- Landslide inventory and hazard risk zonation
- Multi-hazard analyses and multi-hazard modeling
- Identification of catchment stabilization measures

River Hydrology and Morphology Studies
- Hydrological data analyses and hydrological modeling
- River modeling, river hydraulics, and sediment transport analyses
- Identification of river morphology stabilization and mitigation measures

Studies on Headworks Structures
- Damage assessment of civil structures, hydromechanical installations, and instrumentation
- Identification of headworks rehabilitation, remediation, and/or reconstruction measures

This report summarizes Phase 1, analyzes the river catchment, and outlines a broad spectrum of measures, including mitigation, new designs, and the rollout of early warning systems (EWS) that can reduce disaster risks and minimize interruptions to Kathmandu's water supply.

2. FIELD SURVEYS AND DATA COLLECTION

Data and information required for the Phase 1 studies were collected from site visits, various field surveys and investigations, laboratory tests, model studies, and available literature.

Site Reconnaissance Visits

During the study, ADB's TA team of engineers, geologists, a river morphologist, and a hydrologist, undertook extensive reconnaissance of the Melamchi River and its catchment area by helicopter, roads, and foot. Access to these areas, especially those affected by landslides and riverbank erosion, was difficult because of the steep and narrow river valleys and gorges. Nevertheless, the visits provided the team with a valuable opportunity to understand the causes, nature, and scale of the 2021 disaster and to collect data and information through visual inspections and discussions with local residents.

In its initial visits, the team flew by helicopter along the Melamchi River course upstream of the headworks, its tributaries, and potential sources of glacial flows in the area. It also landed at several places en route, including the glacial lake that was the source of debris floods, the Bhemathan deposit, and the headworks site. The team documented its findings through photographs, videos, and notes.

Technical Assistance team members talk to villagers in Sarkathali during a reconnaissance visit (photo by Technical Assistance team).

Subsequent visits to the region covered the upper catchment and the headworks, including its downstream areas. These visits, conducted on foot or by air, yielded useful information about the river morphology, catchment geology, land instabilities, and bank erosion. They also offered the team greater insight into the sediment transport, erosion, and deposition phenomena as it sought to understand the determinants and impacts of the 2021 debris floods.

Images captured by the TA team clearly showed the impacts of the flood events in key areas of the Melamchi catchment including the gently sloping Bhemathan plain. The pre- and post-flood images below show fresh sediment deposited on the dense vegetation of Bhemathan after the first event and washed away by subsequent floods. The images show the knickpoint[9] eroded upstream after the second flood.

Figure 2: Aerial Images of Flood Impacts at the Bhemathan Deposit

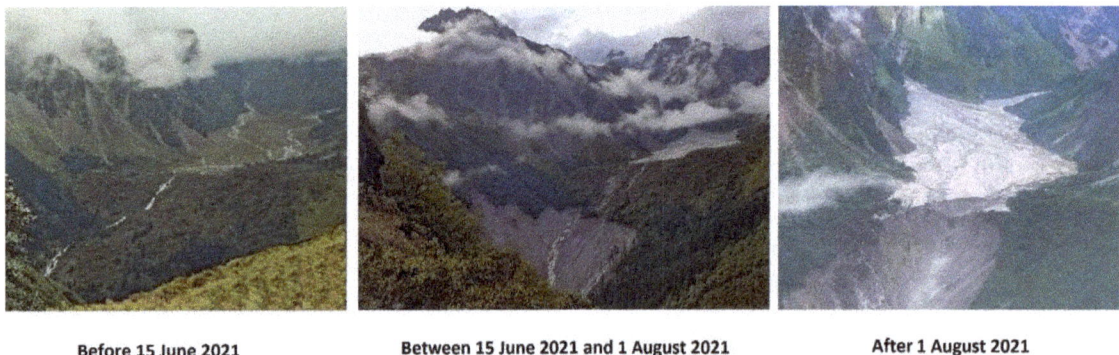

| Before 15 June 2021 | Between 15 June 2021 and 1 August 2021 | After 1 August 2021 |

Source: Authors.

Field Surveys

Topographical Surveys

Topographical surveys of the Melamchi River between the headworks and Bhemathan deposit were performed to generate data for mathematical models and create a baseline for long-term monitoring of the river morphology. These included control surveys to establish permanent benchmarks in the survey area, a longitudinal profile survey of the river, and detailed cross-sectional surveys. Only about 6 km of the planned 14 km survey could be completed because of very difficult access conditions and equipment malfunction due to snowfall and extremely cold weather.

---

9   A hardpoint, such as bedrock or consolidated sediment, which controls the longitudinal shape of the river profile.

Survey work is carried out in the debris-strewn headworks area
(photo by Explorer Geophysical Consultants Pvt. Ltd).

After the surveys, a topographic map of the surveyed area, along with a longitudinal profile and cross-sections of the river, was prepared at suitable scales. Superposition of this map and orthophoto maps[10] of the area revealed some discrepancies between the two, likely caused by changes in the river course after the 2021 floods, plotting inaccuracies due to lack of adequate survey data in inaccessible areas, and equipment malfunction.

Due to the apparent errors in the topographic survey data, digital elevation models (DEM) from drone surveys conducted by Nepal's National Disaster Risk Reduction and Management Authority (NDRRMA) and the World Bank[11] were used for numerical modeling. Notwithstanding discrepancies with the drone maps, the topographic survey map, profiles, and sections will serve as a useful baseline for future surveys to evaluate erosions and depositions in the riverbed.

Bed Sediment Survey

Bed sediment sampling was conducted at three locations to establish the parameters needed for sediment modeling and engineering assessments. In all, nine representative bed samples, three each from the Bhemathan knickpoint area, the headworks area, and areas downstream of

---

10   Aerial drone photographs are corrected geometrically and properly referenced to maps.

11   NDRRMA/World Bank Study. 2021. *Final Report on Aerial Image Acquisition Using Drone Flights of Melamchi, Helambu and Panchpokhari Area (Indrawati and Melamchi River) of Sindhupalchowk.* By consultant, Trimax IT Infrastructure & Services Pvt Ltd.

it, were extracted from near-surface layers of 1 x 1 x 1 m pits, and various field and laboratory tests were conducted on them.[12]

Bed sediment sampling in plots (left) and at Bhemathan (right) (photos by Clean Energy Consultants Pvt. Ltd.)

Tests for particle size distribution (PSD) of the sediment samples were conducted in the field and laboratory. The PSD of large sediments was established by measuring and sieving them in the field, while the PSD of finer materials was analyzed by sieving and hydrometer analysis. A typical PSD curve for sediment samples from the headworks area is shown in Figure 8.

Figure 3: Particle Size Distribution of Sediment Samples from Headworks Area

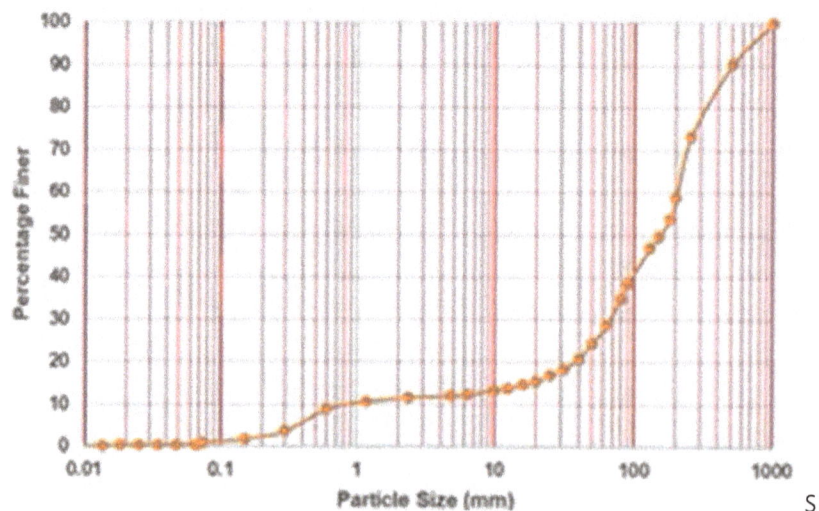

Source: Clean Energy Consultants Pvt. Ltd.

Laboratory tests were also conducted on all samples to determine their moisture content and specific gravity. In addition, direct shear tests were conducted on three samples (<4.75 mm) each from the Bhemathan deposit and the landslide upstream of the headworks. These tests

12   Clean Energy Consultants Pvt. Ltd. 2022. *Melamchi Bed Sediment Survey Report*. Subcontractor to TA 6596-NEP for ADB and MWSDB.

yielded angles of friction of around 30° and cohesions of 0 to 1.5 kilonewtons per square meter (kN/m²) for the Bhemathan samples and 30° and 1.5 kN/m², respectively, for the samples from the headworks landslide.

Geophysical Surveys

Geophysical surveys at Bhemathan and the headworks landslide were carried out to establish their subsurface conditions, estimate the quantity of erodible sediments at Bhemathan, and assess the risk of their erosion. The surveys consisted of two-dimensional electrical resistivity tomography (ERT) and seismic refraction tomography (SRT) along selected profile lines. At Bhemathan, eight ERT and 36 SRT profiles, with aggregated lengths of 3.82 km and 3.45 km, respectively, were carried out. However, three ERT profiles totaling about 1 km produced data errors because of cable faults caused by snow cover and low temperatures. At the headworks landslide, five ERT profiles and six SRT profiles, totaling 1,865 m and 690 m, respectively, were surveyed.

Geophysical sampling preparation on the landslide slope near the Melamchi headworks, February 2022 (photo by Explorer Geophysical Consultants, Pvt. Ltd.).

The geophysical surveys of the Bhemathan deposit show four to five layers of soil and rock materials.[13] Its top thin layer, averaging 3-5 m in thickness, consists of dry and loose recent deposits of coarse sediments, mostly gravels with medium to large boulders toward its upstream, and fine sediments toward its downstream (Figure 4). The second layer is an old deposit that consists of compact fine sediment deposits with thickness between 10 m and 30 m. Fractured bedrocks of gneiss are found at shallow depths of around 15 m below toward the upstream side of the deposit, but at greater depths of around 35 m toward its downstream. The average thickness of fractured bedrock is 5-10 m. Hard and competent bedrock of gneiss is found from depths varying between 20 m and 40 m from the deposit surface.

---

13   Explorer Geophysical Consultants Pvt. Ltd. 2022. *Report on Geophysical Investigation using Electrical Resistivity Tomography Survey (2D-ERT) and Seismic Refraction Tomography Survey (2D-SRT) at the Bhemathan Deposit.* Subcontractor to TA 6596-NEP for ADB and MWSDB.

Figure 4: Seismic Refraction Tomography Survey in Bhemathan

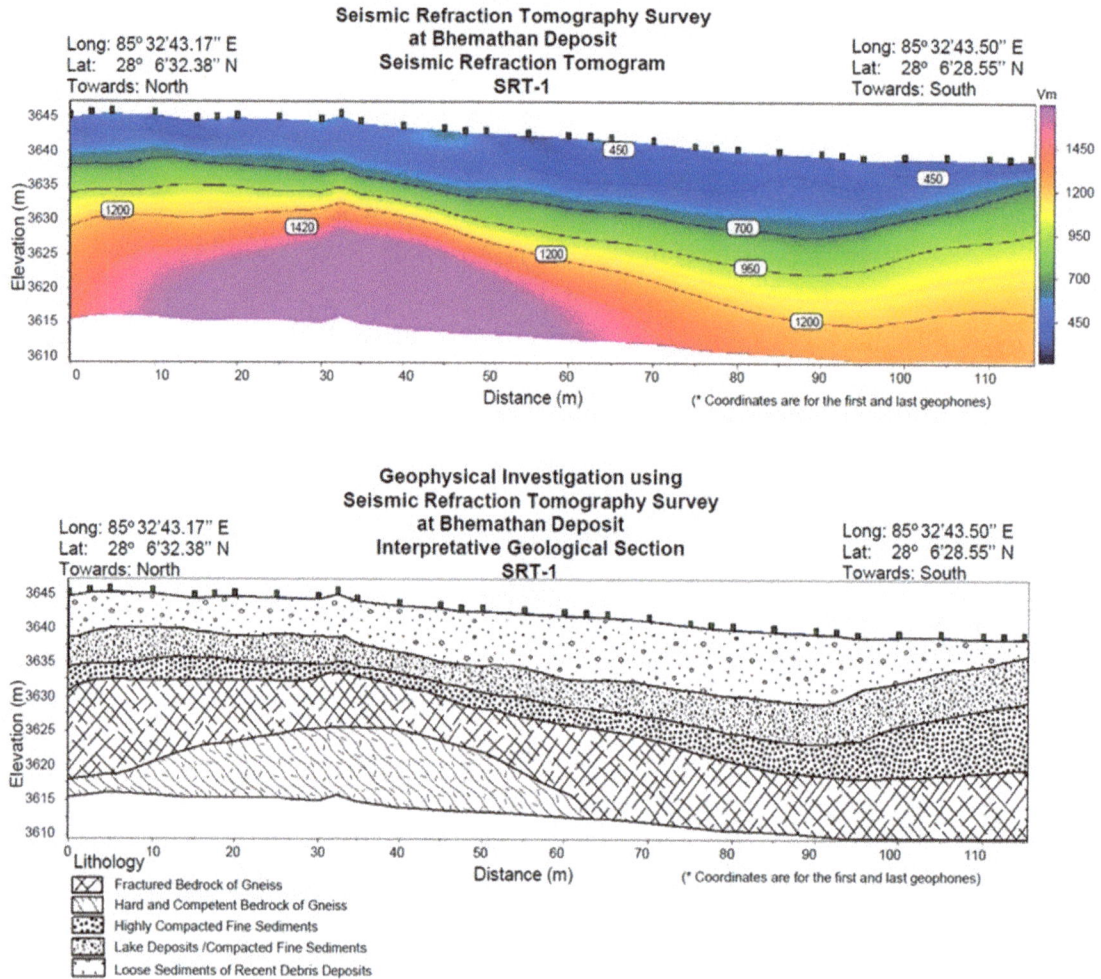

Seismic Refraction Tomography Survey at Bhemathan Deposit
Seismic Refraction Tomogram
SRT-1

Long: 85° 32'43.17" E
Lat: 28° 6'32.38" N
Towards: North

Long: 85° 32'43.50" E
Lat: 28° 6'28.55" N
Towards: South

(* Coordinates are for the first and last geophones)

Geophysical Investigation using
Seismic Refraction Tomography Survey
at Bhemathan Deposit
Interpretative Geological Section
SRT-1

Long: 85° 32'43.17" E
Lat: 28° 6'32.38" N
Towards: North

Long: 85° 32'43.50" E
Lat: 28° 6'28.55" N
Towards: South

Lithology

- Fractured Bedrock of Gneiss
- Hard and Competent Bedrock of Gneiss
- Highly Compacted Fine Sediments
- Lake Deposits /Compacted Fine Sediments
- Loose Sediments of Recent Debris Deposits

(* Coordinates are for the first and last geophones)

m = meter, SRT = seismic refraction tomography, vm = velocity model.

Source: Explorer Geophysical Consultants Pvt. Ltd.

The geophysical surveys at Bhemathan show that the average thickness of the deposit is 25 m. Accordingly, the total volume of the deposit at this location is estimated at 21 million m³ (cubic meters).

At the headworks landslide, the geophysical surveys show the presence of three to four layers of materials. These consist mainly of a top dry landslide mass/colluvial deposit/loose sediment deposit with boulders and fractured rock fragments, fine sediment/compact sediment deposit, and weathered/fractured to hard/competent gneiss. Bedrock is found at shallow depths of 10-12 m toward the toe of the landslide, but at greater depths of 15-25 m toward the crown. Fractured bedrocks extend to greater depths in the southern part of the landslide body.

The slip surface of the landslide is found at varying depths. Its depth is estimated at 15-25 m in the crown, 10-15 m in the main body, and 8-10 m toward the toe.

Engineering Geological Surveys

Engineering geological surveys of the Melamchi catchment up to the headworks, including the headworks landslide, were conducted to collect geological and geotechnical data required for the multi-hazard analysis of the catchment and for design of land stabilization measures. The surveys consisted of (a) engineering geological mapping of relevant areas of the catchment, particularly those with land instabilities, to collect information on geology and geomorphology; (b) field tests to estimate the geotechnical properties of land instabilities; and (c) field studies to collect information on permafrost, rock terrain, etc. The surveys focused on creating enough heterogeneity to the data collected within the catchment.

During the survey, landslide data were collected using pre-designed templates and tables. For each landslide, field soil classification was conducted, and index properties of soils and rocks were estimated. Simple global positioning system surveys were conducted to evaluate each landslide.

Field data for the multi-hazard analyses were collected from the lower catchment through soil classification and inverse borehole infiltration tests. For the upper catchment areas, expert judgment was used to evaluate the soils and rocks.

Results of the geological surveys were used with results from the other surveys to characterize the river catchment. They were also used to prepare a landslide inventory and hazard zonation of the catchment as well as to conduct detailed studies of landslides at the headworks and along the river corridor. In addition, information from the surveys was used with results of hydrological and river morphological studies for a multi-hazard risk analysis of the catchment.

Literature Review

Studies by World Bank and National Disaster Risk Reduction and Management Authority (NDRRMA)

After the 15 June 2021 flood, a technical team established by the World Bank and National Disaster Risk Reduction and Management Authority (NDRRMA) carried out three studies on the Melamchi and surrounding catchments. The first of these was focused on detection of land displacement in the Melamchi, Yangri, and Larke river catchments using Earth Observation remote sensing.[14] The second study used high-quality drone imagery survey, including image and DEM analysis, to collect information on habitation damage, sediment deposition volumes, damage to land use, and changes in the river profile pattern after the floods. In the third study, flood modeling of the Melamchi River was carried out to simulate the flood depth, velocity, and other parameters, and to delineate flood hazard zones.

---

14   Pamungkas, B. et al 2021. *InSAR-based Slope Susceptibility Characterization in Melamchi River Upstream Area.* Synspective Inc.

The World Bank/NDRRMA studies provided useful information on the susceptibility of slopes to landslides in the Melamchi catchment and its zones of highest intensity displacements. They provided insights into the susceptibility of three areas that required further attention: the Pemdan River in the upper Melamchi catchment, a catchment in the north of the Yangri River, and the western edge of the Melamchi watershed.

The DEM developed in the World Bank/NDRRMA study was extensively used by ADB's TA team in Phase 1 modeling works and for landslide and other geological and geomorphological analyses. Local sediment volume estimates developed in the study were also used for the sediment transport calculations in Phase 1. Likewise, results of the flood modeling were extensively by ADB's TA team to optimize its model setup.

Studies by ICIMOD

In autumn 2021, scientists from the International Centre for Integrated Mountain Development (ICIMOD) visited Bhemathan and the Yangri River to inspect the impacts of the landslides. The scientists reported that while the source areas of debris are located below the snow line in October and the 4,000 m contour line, all major deposits were between 3,000 m and 4,000 m, which corresponds largely with the permafrost transition zone. ICIMOD concluded that most mass movements were sourced in a relatively narrow elevation band that corresponds to the area where (a) the state of permafrost is discontinuous or uncertain, (b) precipitation falls in liquid form in the shoulder seasons, and (c) snowmelt is a likely important part of the water balance.

ICIMOD's findings provided vital information on the source areas of debris and its deposition zones. Its correlation of the debris sources with hypsometric and meteorological features also guided the TA team's studies.

Previous Hydrological Studies in the Melamchi River Catchment

Over the years, many researchers have studied the Melamchi River basin but only a few have used statistical analysis and water accounting methods.[15] Most have instead largely used field survey and statistical methods to assess its environmental, institutional, and socioeconomic aspects.[16]

---

15   Mishra, V. S. 2000. *Integrated Development and Management of Water Resources for Productive and Equitable Use: Water Accounting for Indrawati River Basin, Nepal.* Water and Energy Commission Secretariat (WECS) and International Water Management Institute (IWMI).

16   IUCN. 1999. Melamchi Diversion Scheme Environmental Impacts Assessment (EIA) Report. International Union for Nature Conservation (IUCN), Kathmandu, Nepal.

Among the previous studies, a paper published by Chiranjivi Sharma[17] in 2002 contains information relevant to Phase 1. The paper highlights the massive impact of high rainfall on soil erosion and landslides in the Indrawati River basin[18] and discusses the challenges posed by construction works on the basin's dynamic ecosystem.

Disaster Risk Studies in Himalayan River Basins by ICIMOD

One of the primary resource hubs is ICIMOD, which serves as the regional intergovernmental learning and knowledge sharing center for Nepal and seven other member nations. It has produced numerous reports based on high-quality research which have been used by the TA team. Its publications related to GLOF, glacial lakes, landslides, and floods are available online at https://lib.icimod.org/.

---

17   Sharma, C. 2002. *Effect of Melamchi Water Supply Project on Soil and Water Conservation in the Indrawati River Basin, Nepal.* Paper presented at 12th ISCO Conference, Beijing.

18   The Melamchi River is a major tributary of the Indrawati River.

3. MELAMCHI CATCHMENT ANALYSES

The Melamchi River catchment lies in the Himalayan region of Nepal, encompassing pristine glacial lakes, sharp-crested ridges, precipitous gorges, and steep river valleys. Its elevations range from 5,916 meters above sea level (masl) in the north to 809 masl at the confluence of the Melamchi and Indrawati rivers at Melamchi Bazaar.

In its upper reaches, the Melamchi River flows through a narrow valley with gradients as high as 17%. Its riverbed flattens to 8% in the headworks area, from where the river channel widens considerably. With the change in its gradient, the river transits from an erosional channel in its upper reaches to a depositional channel at and downstream of the headworks.

The steep river gradient in the upper reaches of the catchment is broken by the Bhemathan plain located 16 km upstream of the headworks. This 2 km long, 500 m wide plain, formed with accumulated sediments from upstream tributaries over many centuries, has a knickpoint at its downstream end. The knickpoint, possibly formed by an ancient landslide or rockfall, marks the transition from the nearly horizontal river flow over the Bhemathan basin to a much steeper flow immediately downstream.

Erosion of the knickpoint during the 2021 flood events led to massive outflows of water and sediment from the Bhemathan plain and created an amphitheater-shaped scour at its end. These sediments, along with those eroded by the floods in the reach between Bhemathan and the headworks, were deposited in the headworks area and further downstream.

The narrow upstream reach of the Melamchi River is in bedrock, but its reach downstream of the headworks area is an alluvial channel. At Nakote, 4.5 km upstream of the headworks, extensive bedrock erosion during the 2021 floods has caused the river to flow more than 10 m below its previous Bed-Level. This indicates the challenge of situating instream headworks in the river.

The headworks area is situated in a narrow valley with a dissected ridge on the right bank and a reactivated deep-seated gravitational slope deformation (DGSD) landslide on the left, posing a future risk to the headworks structures. Landslides are common in the area.

Regional Geology of the Catchment

Geologically, the Melamchi River basin is situated in the Higher Himalaya Zone of the Bagmati-Gosainkund Region.[19] The area north of Kathmandu is described as a Gosainkund tectonic bridge which has survived erosion from two major rivers: Trishuli on the west and Sun Koshi on the east.

---

19   Dhital, MR. 2015. *Geology of the Nepal Himalaya*. Springer International Publishing, Switzerland.

This region contains a large doubly plunging mega fold, called the Mahabharat Synclinorium, to its south.[20]

Figure 5: Regional Geological Map of Bagmati-Gosainkund Region

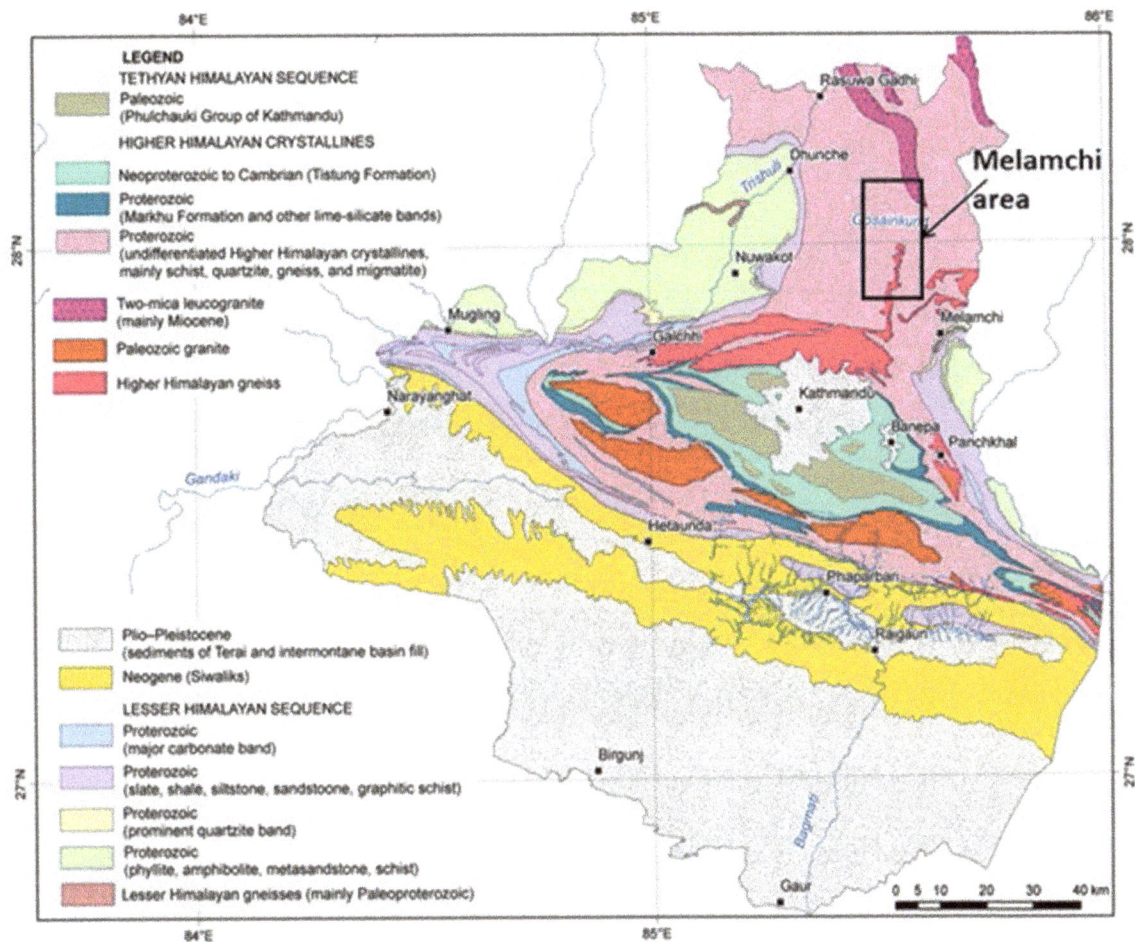

Source: Dhital, M.R. 2015. Geology of the Nepal Himalaya. Springer International Publishing, Switzerland.

The Melamchi River basin lies in the undifferentiated Higher Himalayan crystallines, mainly schist, quartzite, gneiss, and migmatite. Higher Himalayan gneiss is also distributed in the area along with leucogranite in the upper catchment area.

In the Melamchi catchment, the Higher Himalayan crystallines contain large masses of augen and banded gneisses and possess a relatively higher grade of metamorphism compared to similar geological units of the southern area. The rocks comprise the northeast limb of the Mahabharat Synclinorium and part of the Gosainkund tectonic bridge. Based on lithological characteristics,

---

20   Hagen, T. 1969. *Report on the Geological Survey of Nepal. Volume 1. Preliminary Reconnaissance.* OCLC Number/Unique Identifier: 898846648.

the rocks are classified into five geological formations: Talamarang Formation, Gyalthum Formation, Bolde Quartzite, Timbu Formation, and Golphu Formation.[21]

The **Talamarang Formation** has light gray to dark gray kyanite–garnet–biotite schists, banded gneisses, and quartzites. A few bands of white or pale green, clean or sporadically calcareous quartzite also occur together with sillimanite gneiss or schist. This formation is near the Main Central Thrust (MCT) zone which passes through nearly 5 km downstream from the Melamchi–Indrawati confluence at Majhitar in the Sindhu Khola area.

The garnet–chlorite schists of the MCT zone are followed upward by the kyanite–garnet schists into the Talamarang Formation. The MCT zone is characterized by a thick zone of garnet–chlorite schist that rests over fine-grained white quartzite band underlain by the Benighat Slates of Nawakot Complex of Lesser Himalayan Zone.

The **Gyalthum Formation** consists of more than 80% thin- to thick-banded, light gray to gray, laminated quartzites with mica partings, feldspar–garnet–biotite schists and bands of augen gneisses. The quartzites are laminated and intensely deformed. The Gyalthum Formation transitionally passes into the overlying Bolde Quartzite.

The **Bolde Quartzite** consists of thick to very thick, massive, medium- to fine-grained quartzite bands with mica partings. A few sillimanite-bearing thin schist, bands also occur in the Bolde Quartzite. The upper part of the Bolde Quartzite unnoticeably grades into the overlying Timbu Formation. It is found as a thin formation in the Melamchi catchment.

The **Timbu Formation** is composed of light to dark gray, intensely deformed quartzites, schists, banded gneisses, and migmatites with abundant sillimanite. It is a discontinuous unit lying between the Bolde Quartzite at its base and the Golphu Formation at its top. The migmatite zone mainly occupies the central and upper levels of the formation and exhibits various types of ptygmatic folds, small-scale faults, and flow structures. It can be well observed in the headworks area and its uphill sections.

Many quartzite bands occur in the Timbu Formation. Banded gneisses are also found in lower parts of the Timbu Formation.

The Timbu Formation in the headworks area has a dip slope valley and, as a result, most of the left bank of the Melamchi River has large-scale landslide deposits and active landslides along with DGSDs.

The **Golphu Formation** has pegmatite intrusions and many augen gneiss bodies within the formations. It contains coarse-grained, thick-banded, dark gray feldspar–garnet–biotite schists

21   Dhital, M.R. et al. 2002. Geology and structure of the Sundarijal–Melamchi area, central Nepal. *Journal of Nepal Geological Society*, vol. 27.

and banded gneisses, with a few laminated quartzite bands. This formation is crosscut by many pegmatite veins. A remarkable pegmatite body is found in the right-side hill section of the headworks area. There are also a few large zones of augen gneiss in this formation. Nearly half of the catchment has Golphu Formation with many quartzites, augen gneiss, marble bands, calcareous gneiss, and pegmatite bands which are not in mappable units in the catchment scale mapping.

The Golphu Formation also has a dip slope in the left bank, and landslide deposits are common in this formation. In fact, the right bank of the Melamchi valley slope is a counter dip slope and mainly has rockfall problems. A few talus deposits are also noticed in the valley near the Melamchi Ghyang village.

Figure 6: Geological Map of the Melamchi Catchment

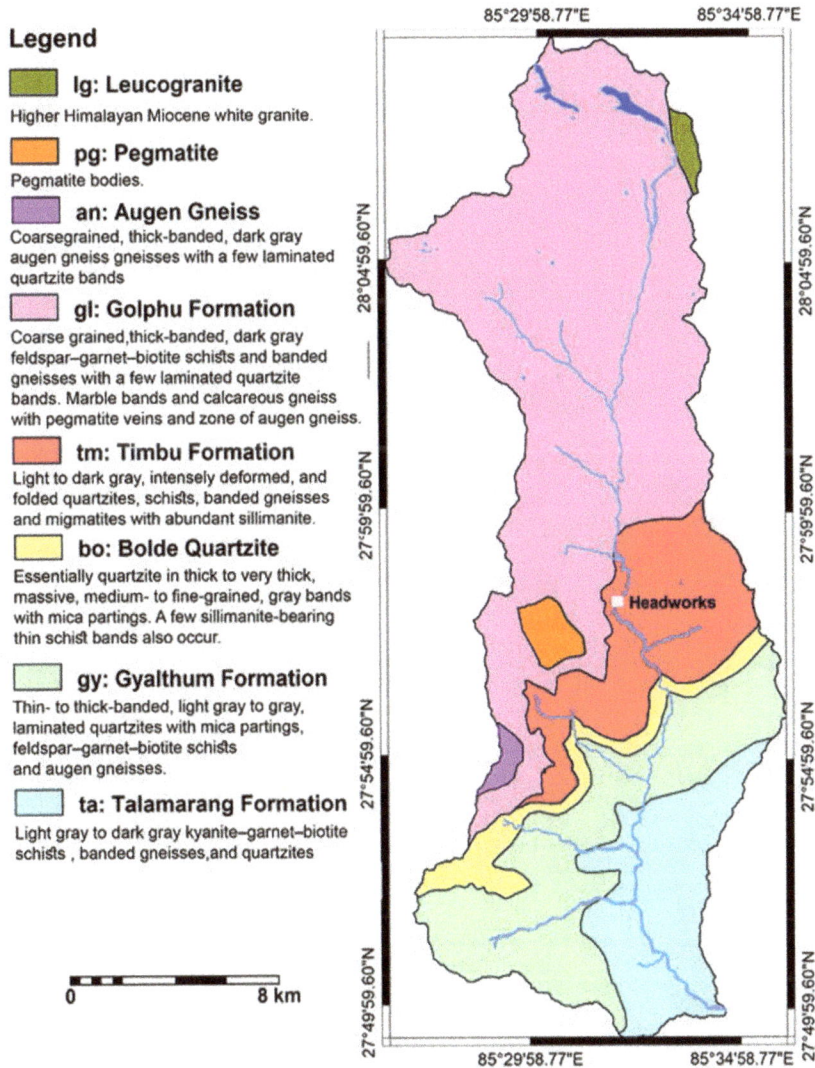

Source: Authors using field data and information from Dhital, M.R., 2015. Geology of the Nepal Himalaya. Springer International Publishing, Switzerland.

Geomorphology

Geomorphological maps of the upper Melamchi catchment were prepared from field data and remote sensing data considering 20 terrain units identified during the field surveys. For permafrost areas, a simple permafrost delineation was assumed along the 4,200 m contour, and the surface features of the higher elevation were mapped through videos and photos taken during a 5- hour helicopter flight.

Analysis of the distribution of the various terrain units in the upper catchment shows that the catchment had more than 25% non-forested (bare) rocky terrain and over 14% deglaciated terrain and permafrost. The geomorphological maps show that soil-related issues mainly prevail at lower elevations between Melamchi Ghyang and the headworks, whereas rockfall and rockslide problems mainly exist in the rocky terrain in the higher elevation area between the two sites. The mapping also showed that the Bhemathan area exhibited extreme erosion and bank cutting.

Figure 7: Geomorphological Map of Melamchi River Upper Catchment

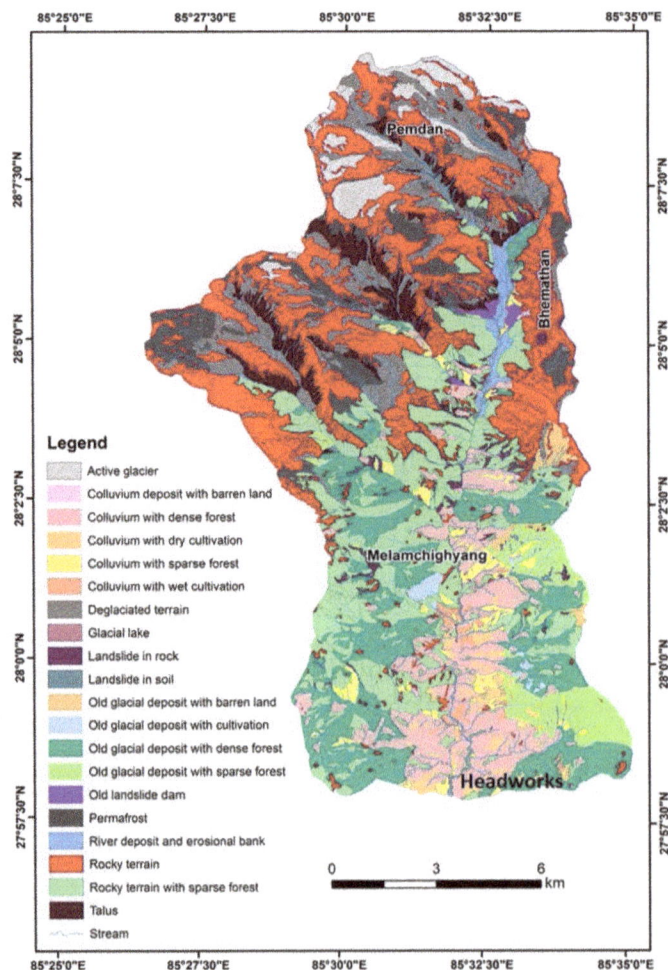

Source: Authors using field data and information from Dhital, M.R. 2015.
Geology of the Nepal Himalaya. Springer International Publishing, Switzerland.

The left bank of the Melamchi River upstream of the Melamchi Ghyang mainly consists of dip slope[22] in rocky terrain. Another dip slope is present further downstream, but it is mostly covered with colluvium and old glacial deposits at elevations higher than 2,200 m. Such glacial deposits are seen at Tarke Ghyang, from where the downslope to Nakote is covered with colluvium.

Typical dip slopes on the entire left bank of the upstream stretch of the Melamchi River make the river channel narrow due to slow-moving creep along the foliation joint that daylight on the slope. These creeps have formed colluvium on the slopes and are known as DGSDs or old landslide deposits. They are porous and have a relatively lower drainage density than the surrounding rocky terrains. At lower elevations, these colluvium deposits have permanent springs and seepages.

Another serious issue of headward erosion and possible GLOF-related flash floods is noticed in the areas around glacial lakes upstream of Namsan Creek. The creek itself has many erosion areas and has rockfall scree with talus deposits[23] on the left bank above the deglaciated valley and permafrost.

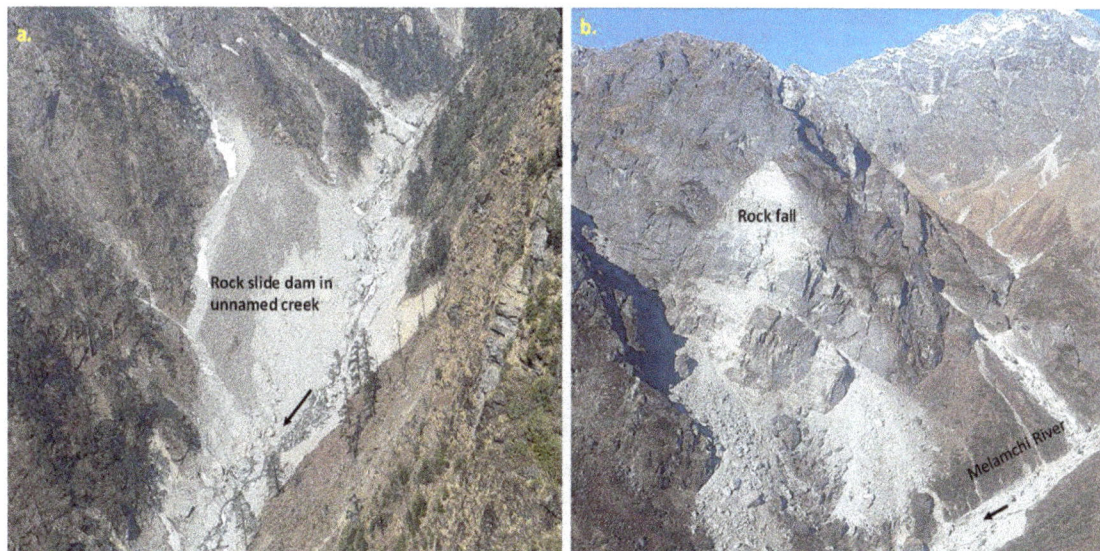

Rockfall and Rockslides are shown in an unnamed creek between Namsan Creek and Bhemathan. Arrow shows debris flow direction (photo by TA team).

22   A topographic surface which slopes in the same direction as the underlying rock.
23   Piles of rocks accumulated at the base of a cliff.

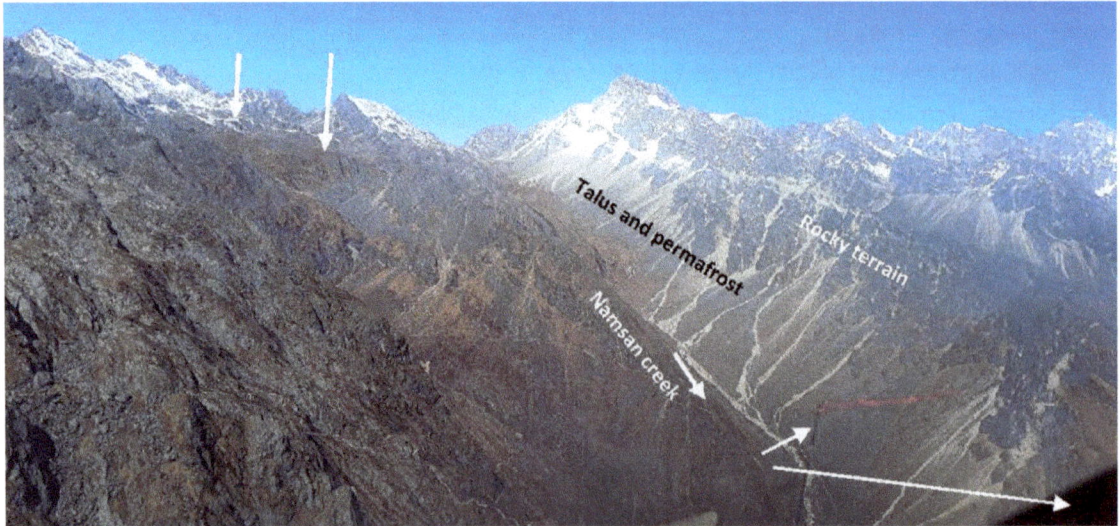

Fragile geomorphological settings of Namsan Creek (photo by TA team).

Headworks Area

The headworks area is influenced by a dissected ridge of banded gneiss which results in a change in the valley alignment from the north-south direction to the north-southeast direction.

This ridge has blocked the river in the past, and continued erosion in the banded gneisses has created a deep gorge in the river. The effect of the ridge can be seen in the valley, where old, stratified reservoir deposits are noticed on the right riverbank.

Figure 8: Downstream View of Gorge Area and Headworks

Source: Authors.

Figure 9: Upstream View of Headworks and Gorge Area

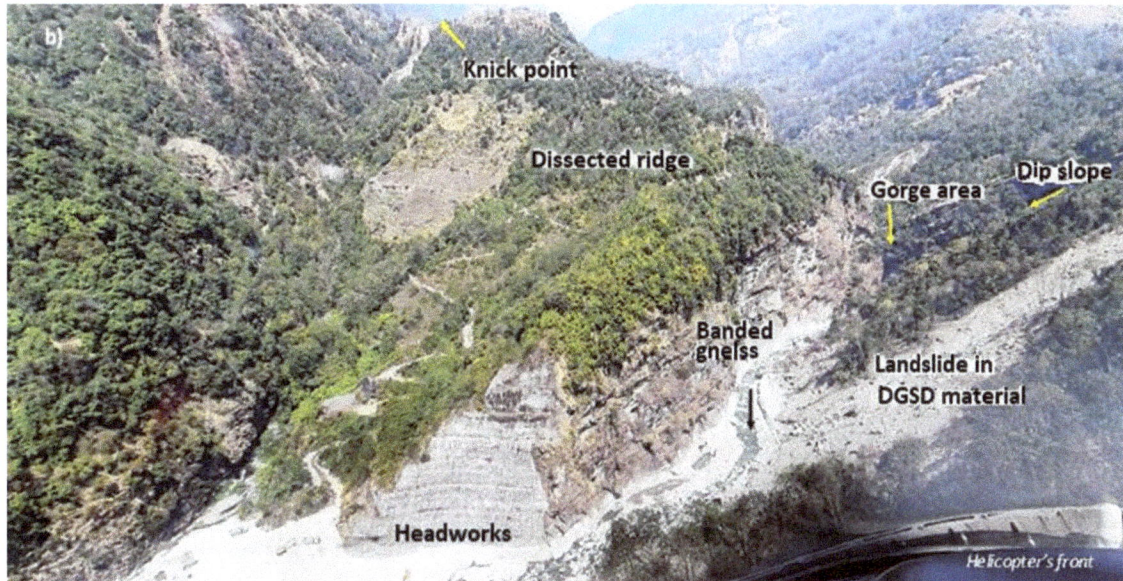

Source: Authors.

Some old reservoir deposits are also observed immediately upstream of the headworks area. These old DGSD sediments on the downstream left bank have caused the river to shift its course toward the southwest in the weir area, resulting in sudden changes to the river's sedimentation and flow patterns in this reach compared with the upstream flow. The rock orientation along the left bank of the river creates a dip slope, which has caused the DGSD to reactivate after heavy rainfall and toe cutting during the 2021 floods.

Landslide Distribution

Landslides in the Melamchi catchment are categorized into (a) earthquake-induced landslides, mostly from the 7.8-magnitude 2015 Gorkha earthquake; and (b) rainfall-induced landslides. Accordingly, an inventory map of the landslides in the catchment was prepared by the TA team using a time series of satellite images with earthquake-induced landslide data from the United States Geological Survey, alongside their own database of rainfall-induced landslides.

Figure 10: Melamchi Catchment Landslide Inventory Map

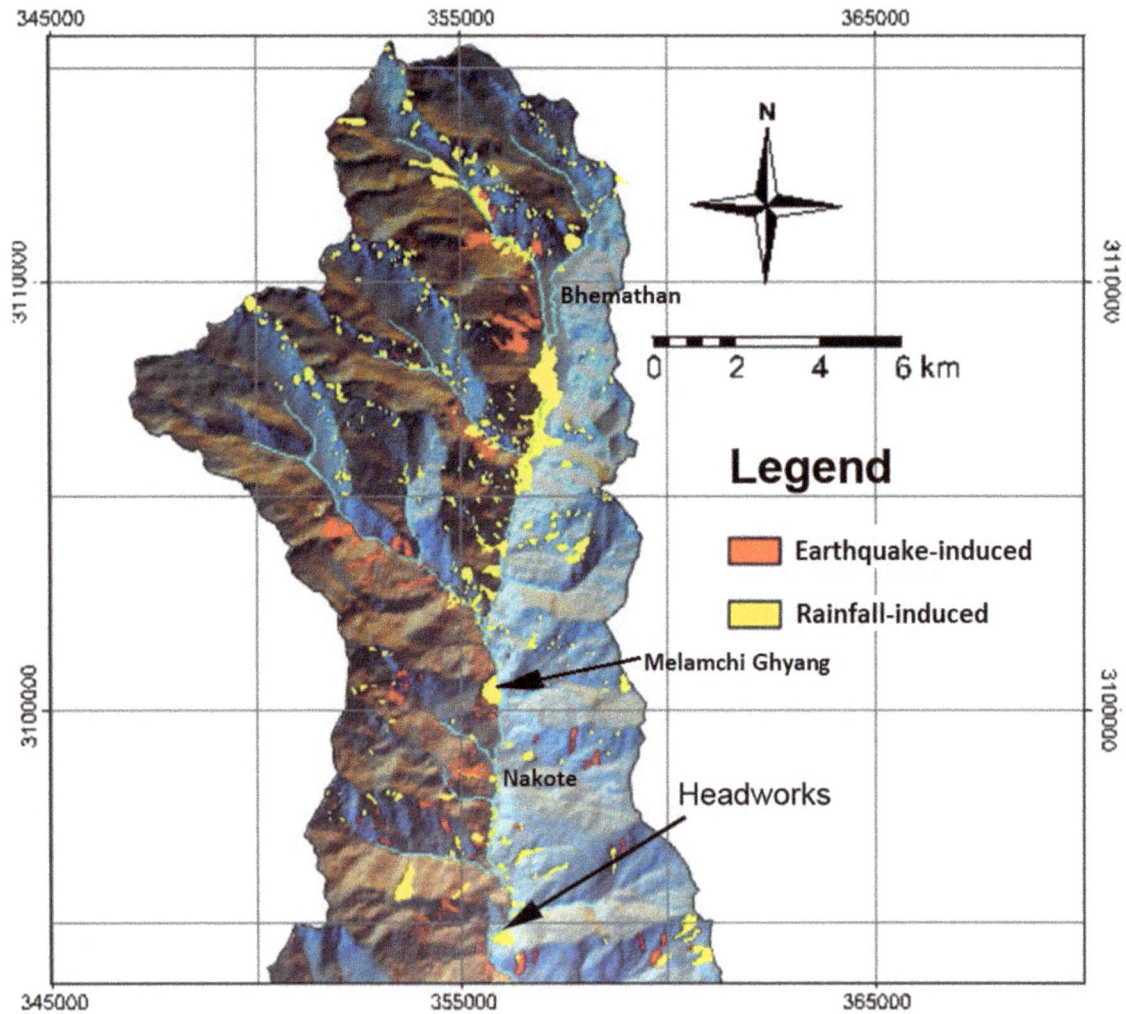

Source: Authors, with data from the United States Geological Survey. Roback, K. et al. U.S. Geological Survey. 2017. https://doi.org/10.5066/F7DZ06F9.

Glacier Distribution Map

The Melamchi catchment is partly covered with glaciers, with deglaciated valleys and old morainal materials in its upper ridges. Ten glacial lakes and at least two glaciers are present in the catchment, but only a few of the glacial lakes pose a threat of sudden outburst for the downstream reaches.

Glacial lakes in the Namsan Khola catchment, located northwest of Melamchi Ghyang, are small and have limited water volumes. These lakes are mostly dammed by bedrock with thin layers of moraine and, therefore, differ from the breached glacial lake in Pemdan Khola that was completely blocked by thick moraine and continuously fed by glacio-fluvial flows. Nevertheless, as observed by the TA team during a helicopter survey, the dams of moraine around each of the six glacial lakes in the Namsan Khola catchment show signs of erosion.

The Namsan Khola joins the Melamchi Khola from the right, upstream of Melamchi Ghyang. This creek itself has many erosion problems and rockfall scree[24] with talus deposits on the left bank that are deposited on the deglaciated valleys and permafrost.

Figure 11: Four Glacial Lakes in the Upstream Melamchi Catchment

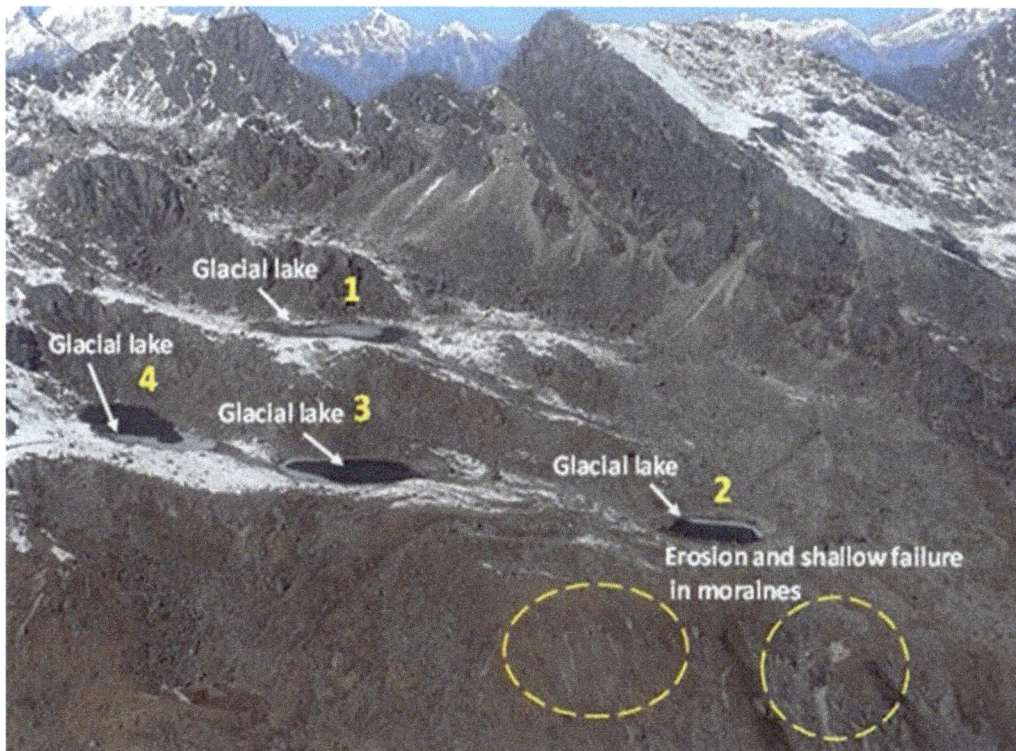

Source: Authors.

---

24   Scree is a collection of broken rock fragments at the base of a cliff.

Figure 12: Glacial Lakes Inventory Map

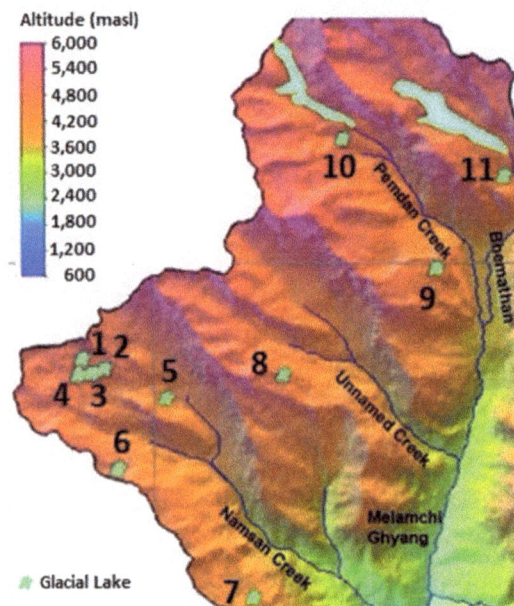

masl = meters above sea level

Source: Authors.

Land Displacement Analyses through SAR Satellites

Land displacement hot spots in various sections of the Melamchi catchment were analyzed by a World Bank study team for NDRRMA using the free-source Synthetic Aperture Radar (SAR) data available from the European Space Agency.[25] This technology is often used to map landslide-prone areas and provide early warning, however, it detects only slow-moving landslides, not sudden land slope failures. Furthermore, it needs SAR data in both ascending and descending orbits to produce good results.

Landslides Caused by 2015 Gorkha Earthquake

The 7.8-magnitude Gorkha earthquake, with its epicenter at Barpak about 80 km west of the Melamchi headworks area, struck on 25 April 2015. The main shock was followed by many aftershocks, consisting of 421 with magnitudes of 4 or more. Four aftershocks had magnitudes greater than 6, and one with magnitude 7.3 hit on 12 May 2015 with its epicenter in the Sindhupalchok district[26] through which the Melamchi River flows.

Apart from other damage, the earthquake activated or reactivated numerous landslides that caused major damage to settlements, infrastructure, and the environment. In the Melamchi

---

25   Pamungkas, B. et al. 2021. *InSAR-based Slope Susceptibility Characterization in Melamchi River Upstream Area.* Synspective Inc.

26   ICIMOD. 2016. *The Impact of Nepal's 2015 Gorkha Earthquake-Induced Geohazards,* ICIMOD Research Report 2016/1.

catchment, the earthquake triggered 12 landslides on the right riverbank between the Bhemathan knickpoint and the headworks, and at least two on the left bank.

The relatively larger impacts of the earthquake on the right bank are attributable to geological and structural conditions related to the geometry of the discontinuity fabric. The foliations in the catchment dip toward the west, i.e., to the right bank that also has a scarp[27] slope with generally small, loose rock bodies and fine material capable of causing a slide. The left bank has a dip slope with less fine material, where slides of larger blocks could occur locally along dipping foliation planes.

The Melamchi floods in June 2021 triggered a few landslides, including the large slides downstream of the Bhemathan knickpoint on both valley sides, the large Melamchi Ghyang slide, and the landslide opposite the headworks. The Melamchi Ghyang slide dammed the river in the 2021 events, and the subsequent breach of this natural dam caused the outburst of the lake that had built up behind it. Studies indicate that this slide was partially caused by the Gorkha earthquake and activated during the 2021 floods, possibly triggered by toe erosion due to high water levels with high velocities in the river channel. The geomorphological shape of Melamchi Ghyang also suggests the slide could have been an old dormant slide. Experience from other similar areas in Nepal suggests that susceptibility to landslides has increased after the 2015 earthquake. This could be because many landslides initiated by the 2015 earthquake were not fully developed and are in a latent state.

Return Period for 2015 Earthquakes

The Gorkha earthquakes of 25 April and 12 May 2015 caused peak ground accelerations between 0.4 g and 0.6 g in Sindhupalchok district,[28] "g" being the acceleration due to gravity. Rough estimates, based on PGAs for the Melamchi valley published by the Global Seismic Hazard Assessment Program and relations proposed by Euro code,[29] show that the return period of an earthquake similar to the Gorkha earthquake, with PGAs of 0.4 g and 0.6 g, is in the range of 200 to 600 years.

Landslide Susceptibility

Considering the hazards from the damming and subsequent flooding caused by landslides, the TA team carried out landslide susceptibility analyses of the Melamchi catchment to identify areas with the potential to slide due to steep slopes, weak geological units, etc., especially during extreme rainfall. The analyses were performed using two approaches: (a) rapid landslide susceptibility mapping based on slope, and (b) multivariate techniques proven to be effective in spatial prediction of landslides with a high degree of accuracy.

---

27   A scarp in geology denotes a slope in the land cutting across the underlying strata.

28   Gautam, D. and Rupakhety, R, 2021. Empirical Seismic Vulnerability Analysis of Infrastructure Systems in Nepal. *Bulletin of Earthquake Engineering*. 19. 10.1007/s10518-021-01219-5.

29   Eurocode 8: *Design of Structures for Earthquake Resistance, Part 1: General Rules, Seismic Actions and Rules for Buildings.*

Rapid Landslide Susceptibility Mapping Based on Slope

For the susceptibility analysis, a slope map of the Melamchi catchment delineating areas with similar slope angles was prepared using drone-based DEM of the post-event topography of the Melamchi valley. Slope inclination ranges of less than 20°, 20°-35°, 35°-45° and greater than 45° were considered for the mapping, focusing on areas near the riverbed that could be affected by bank erosion or scour.

Based on the properties of the slope deposits, slopes with angles between 20° and 45° were found to be typically prone to landslides, the most sensitive of these being slopes with angles between 35° and 45°. Areas in the critical range between 20° and 35° are concentrated in the uppermost parts of the Melamchi valley, i.e., between the Bhemathan knickpoint and 2.8 km further downstream. Areas of similar slope angles are also found locally up to the headworks area, but generally these areas were much smaller than those near Bhemathan. A few areas in the slope range between 35° and 45° are found locally, with the largest ones located on the right bank some 100 m downstream of the Bhemathan knickpoint and the Melamchi Ghyang slide, and on the left bank opposite the headworks area.

As the kinematic behavior of rock slopes steeper than 45° is controlled by the geometry of their discontinuity fabric, rockslides or rockfalls from areas with such slopes are generally likely to be a local phenomenon and will not contribute much debris. Although the occurrence of large-scale rock or landslides in such areas cannot be completely ruled out, it is very unlikely that such events will be triggered only by bank erosion. Their main trigger will be dynamic forces caused by earth tremors.

While small rockfalls may not contribute much to debris generation, they could well clog narrow gorges. These falls are also more likely to be triggered by seismic events than by cloudbursts. Clogged gorges may cause damming of the valley and may eventually lead to LDOF.

Landslide Susceptibility Zonation Map

A landslide susceptibility map for the Melamchi catchment was prepared using multivariate statistical modeling. For this purpose, a landslide inventory map was prepared for both rainfall- and earthquake-induced landslides. A rainfall-induced landslide raster map was used as a factor map, and nine causative factor maps were developed to generate a susceptibility zonation map using logistic regression modeling. The resulting map is shown in Figures 13 and 14.

Figure 13: Some of the Factor Maps Used for Landslide Susceptibility Models

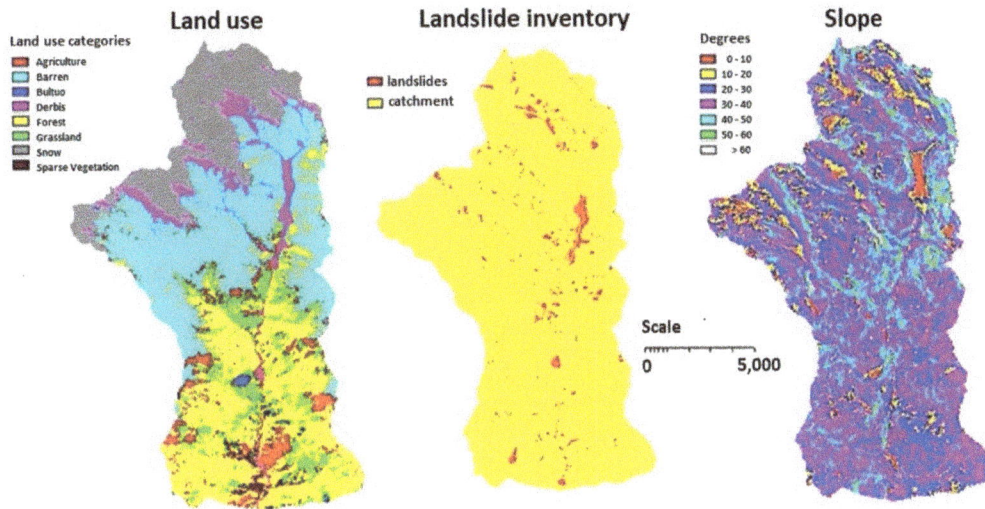

Source: Authors.

Figure 14: Melamchi Catchment Landslide Susceptibility Map

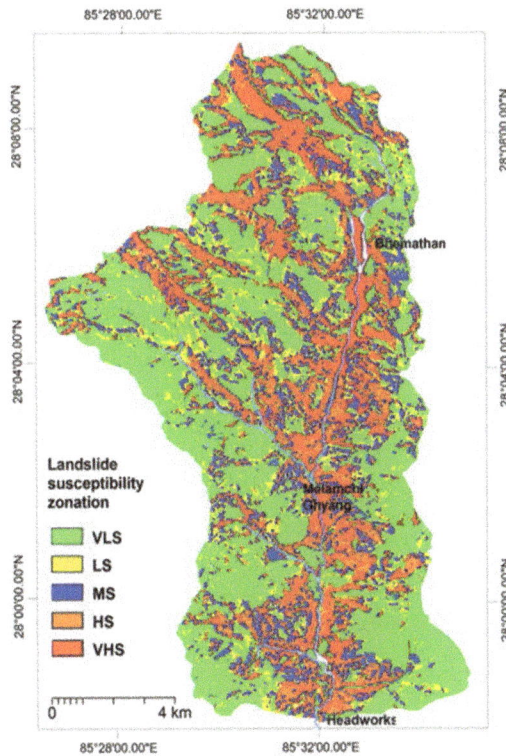

HS = high susceptibility, MS = medium susceptibility, LS = low susceptibility, VHS = very high susceptibility, VLS = very low susceptibility.

Source: Authors.

To understand the power of zonation categories for landslides, the predictions were evaluated by crossing the landslide inventory map. This showed that very high susceptibility (VHS) and high susceptibility (HS) zones cover more than 87% of existing landslides.

The susceptibility maps suggest that landslides hazards will prevail in the Melamchi catchment. Specifically, the area from Melamchi Gyang to Bhemathan is highly susceptible to landslides, a fact confirmed by visual observation during field visits. In addition, almost all tributaries are highly susceptible to landslides.

Summary of Critical Landslide Sites

Combining the results of the analyses, satellite, and drone image studies and field inspections, 15 critical landslide sites that could potentially dam the river or damage nearby structures were identified. Of these, the landslide LS1 upstream of the headworks is deemed the most critical because of its proximity to the headworks, its size (250 m in length along the toe), and the possibility of its further progressing and worsening the headworks condition created by the 2021 floods. The landslide at Melamchi Ghyang (LS15) is also critical owing to its size (more than 700 m along its toe), and the risk of it damming the river again.

Figure 15: Identification of 15 Critical Melamchi River Landslides

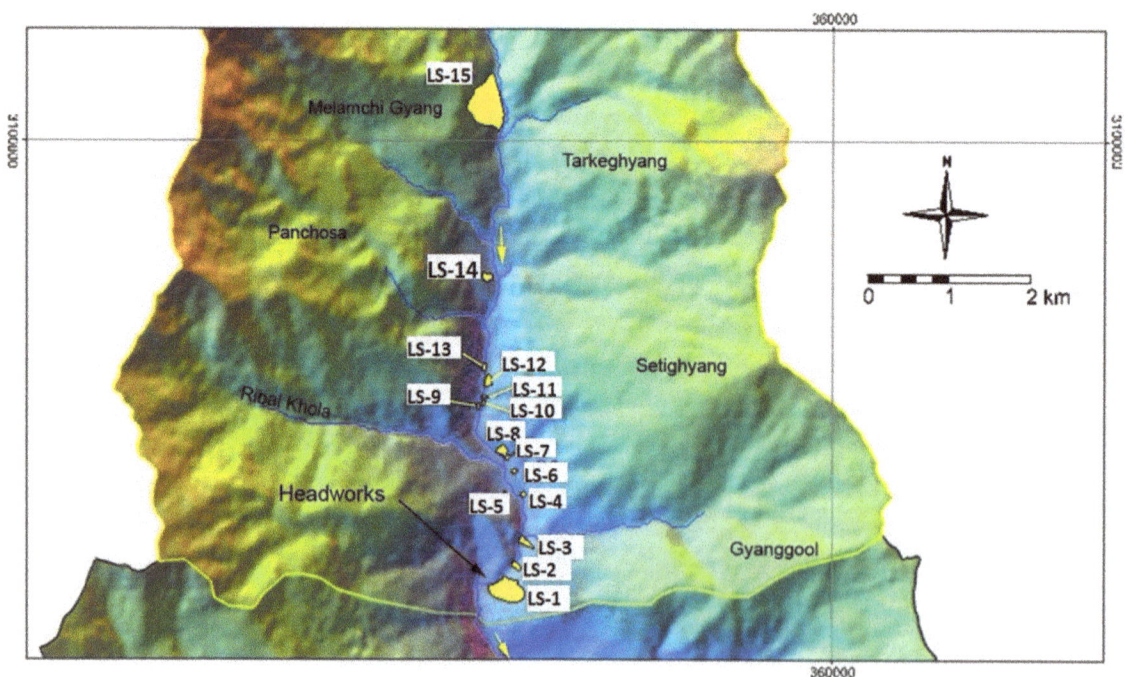

Note: LS1 to LS15 identify individual landslides.

Source: Authors.

All existing landslides in the catchment may affect the Melamchi River directly. In addition, future landslides in the tributaries could travel down their course and dam the river at their confluences.

Landslide Stability Analysis of LS1

A landslide stability analysis was carried out to investigate the risk of continued landslide activity at LS1 on the headworks' left bank. Input parameters for the analysis were adopted from the recent geophysical and geological surveys at this site.

The landslide was analyzed for the dry and wet season scenarios assuming corresponding groundwater tables. The simulation showed the slope near the headworks was stable during the dry season, with a factor of safety well above 1. During the wet season, however, the simulated factor of safety was below 1, implying that the landslide could progress further in the monsoon season, with a high risk of failing suddenly and catastrophically.

Impact of Slope Saturation

As illustrated by the analyses of LS1, the stability of slopes largely depends on their degree of saturation. As increasing moisture content in slope materials reduces their effective shear strength, prolonged or intense rain events, or snowmelting, can increase the susceptibility of the slopes to sliding. In the case of the Melamchi landslides, snowmelt caused by the rapid temperature rise and rainfall over an extended period of time before the June 2021 event contributed to soil saturation, which enhanced landslide susceptibility.

Drainage of slopes is, therefore, vital to reducing or minimizing the water content of slope materials. Alterations to natural or artificial water drainage paths can significantly impact landslide susceptibility. If drains alongside roads that follow slope contours are not properly maintained, or natural surface drainage channels along the slope are blocked, surface water could infiltrate the substrata. This could also occur if artificial or natural drainage of groundwater is blocked at the landslide toe.

4. MELAMCHI RIVER ANALYSES

Hydrological Surveys and Modeling

Hydrological modeling was performed using simulated results from a model calibrated and validated using the Soil and Water Assessment Tool (SWAT[30]) This analysis included water balance, flood frequency, estimation of probable maximum flood (PMF), and the impact of climate change on water balance and hydrological extremes. A separate event-based model was developed using the Hydrologic Engineering Center Hydrologic Modeling System (HEC-HMS)[31] for estimating PMF based on probable maximum precipitation (PMP).

Catchment Characteristics

The total catchment area of the Melamchi River at its confluence with the Indrawati River is 324 km² and at the MRWDS headworks 161 km². The catchment has a very steep gradient up to the headworks, which results in high flow velocities through narrow and deep river sections. It has a dendritic river system.

The catchment elevations vary from 804 m at the confluence with Indrawati River to 1,429 masl at the Melamchi headworks, and then to 5,916 masl. The catchment is covered dominantly by forests and grasslands. Gelic Leptosols (LPi) is the dominant soil type in the entire catchment.

---

30   Arnold, J. G. et al. 1998. Large-area hydrologic modeling and assessment part I: Model development. *Journal of American Water Resources Association,* vol. 34/1.

31   Feldman, A.D. 2000. *Hydrologic Modeling System HEC-HMS, Technical Reference Manual.* U.S. Army Corps of Engineers, Hydrologic Engineering Center, HEC, Davis, California, USA.

Figure 16: Melamchi Catchment Topography and Hydro-Meteorological Stations

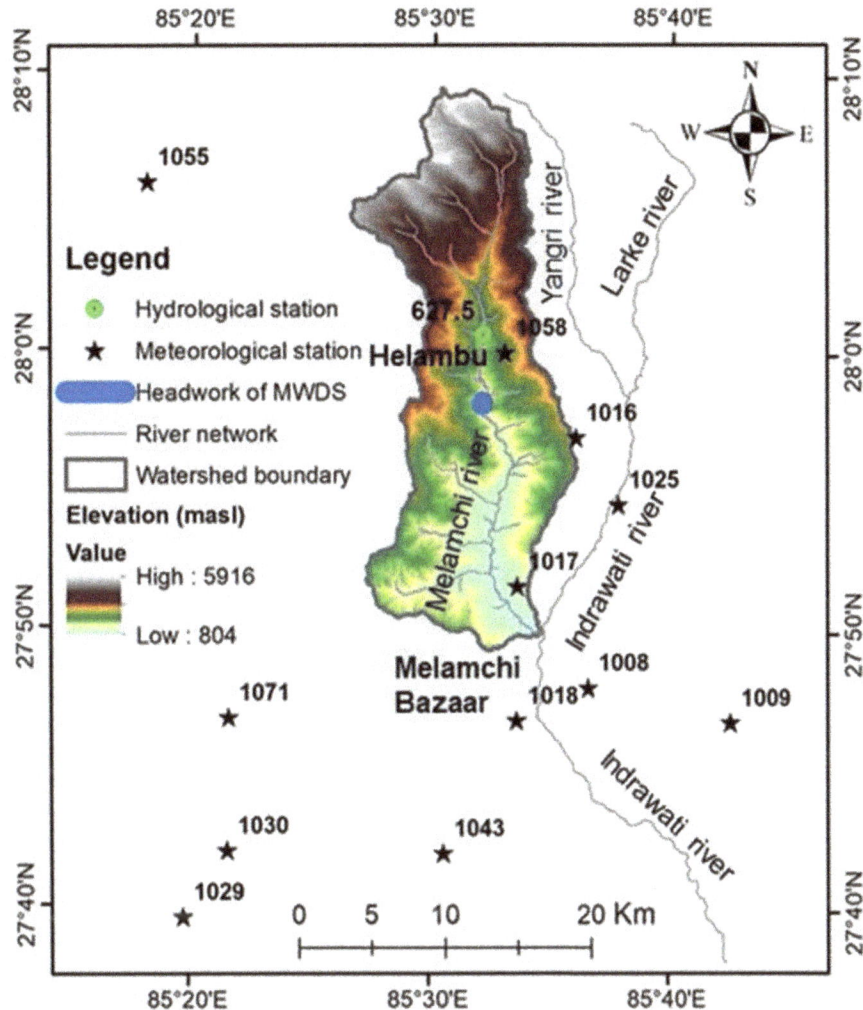

masl = meters above sea level, MRWDS = Melamchi River Water Diversion Subproject.
Source: Authors.

Hydrological Model

The hydrological model for the study was set up using SWAT, which is a basin-scale, spatially distributed parameter and continuous time-simulation model designed to predict the response to natural inputs like precipitation including snowfall and melt, and to human interventions on water management, sediment yields, and agricultural chemicals in ungauged catchments. The model is physically based, computationally efficient, and capable of continuous simulation over long time periods to compute the effects of management changes. It is also able to handle both spatially and temporally variable data as inputs for estimating streamflow through various comprehensive hydrologic processes. The model was, therefore, chosen for this study to capture the spatial variability of input data.

SWAT inputs defined at three levels, namely watershed, sub-basin, and hydrological response unit, were used to model processes throughout the catchment. The model estimated relevant hydrologic components such as evapotranspiration, surface runoff, groundwater flow, and sediment yield for each hydrological response unit. The hydrologic cycle as simulated by SWAT was based on water balance.[32]

For modeling, the Melamchi watershed above the Indrawati confluence was discretized into 29 sub-basins and further into HRUs to capture spatial heterogeneity in the catchment characteristics. The catchment topography was represented in the model by a hydrological-connected DEM of 5 m resolution developed by the Water and Energy Commission Secretariat as part of a river basin planning study. Likewise, soil data was based on the Soil and Terrain database[33] and land use/cover data from ICIMOD publications. In all, five elevation bands were defined to allow variation in precipitation and temperature with elevation. All meteorological inputs were observed data from Department of Hydrology and Meteorology (DHM).

The model was calibrated against observed daily stream flow data at the Helambu hydrological station[34] that has a catchment area of 119 km². The calibration and validation were performed for periods from 1993 to 1996 and from 2001 to 2005, respectively, considering the quality of available streamflow data. Model performances were evaluated based on Nash-Sutcliffe efficiency (NSE), coefficient of determination R^2, and percent bias (PBIAS) as indicators. Furthermore, reproductions of hydrological pattern and flow duration curve were used as criteria for model performance evaluation. Calibration was initially performed automatically using SWAT-CUP with the Sequential Uncertainty Fitting Algorithm (SUFI-2) for parameter optimization. The model was further fine–tuned using manual calibration.

Observed and simulated hydrographs for a calibration period covering full years (all seasons), are shown in Figures 17 and 18. The NSE, R^2, and PBIAS during calibration were 0.78, 0.81 and -6.4%, respectively, and 0.76, 0.79, and -6.0, respectively, during model validation. The performance statistics and hydrograph pattern indicated that model performance was reasonable, and that it could be used for further hydrological analysis. The model could not, however, reproduce the peak in the 1995 post-monsoon due to unavailability of specific data for that event.

The range/order of the annual maximum discharge for the calibration and validation periods can be considered representative for all available observed data. Although much lower than the observed discharge of 2021 (which were larger than 2,000 m³/s), these discharges were essential

32   Neitsch, S. L. et al. 2002. *Soil and Water Assessment Tool: The Theoretical Documentation, Version 2000.* Texas Water Resources Institute, College Station, Texas, TWRI Report TR-191.

33   Dijkshoorn, J.A. and Huting, J.R.M. 2009. *Soil and Terrain (SOTER) Database for Nepal. Report 2009/01.* Accessed 15 December 2016 (Online Dataset).

34   Station 627.5. www.dhm.gov.np/hydology/hydrological-stations

for simulating the rainfall-runoff component of the river flow for instances such as those of August 2021 that were composed of the rainfall-runoff including cloudburst, and dam breaks from natural reservoirs such as glaciers or landslide dams.

Figure 17: Observed and Simulated Hydrograph, Daily Averaged Values, 1993-1996

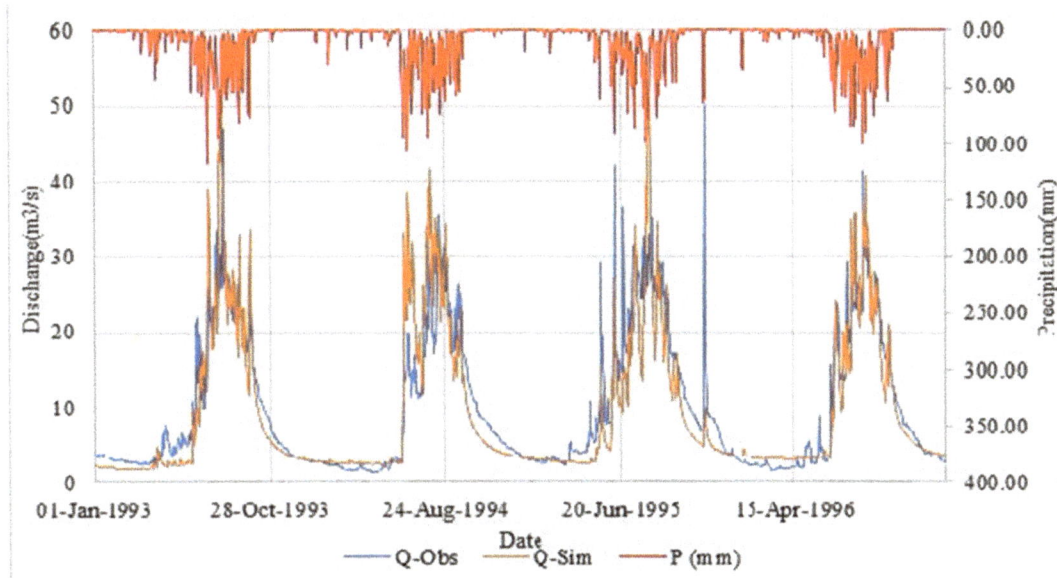

m³/s = cubic meter per second, mm = millimeters, p = precipitation, Q-Obs = observed discharge, Q-Sim = simulated discharge.

Source: Authors.

Figure 18: Observed and Simulated Hydrograph, Monthly Averaged Values, 1993-1996

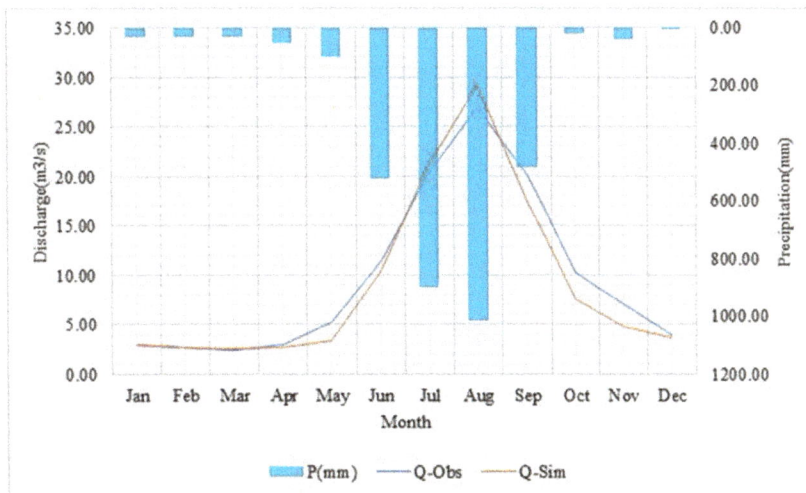

m³/s = cubic meter per second, mm = millimeters, p = precipitation, Q-Obs = observed discharge, Q-Sim = simulated discharge.

Source: Authors.

Hydrological Simulations

Following proper calibration and validation, the SWAT model was used to simulate long-term (1981-2021) stream flows at the MRWDS headworks site. The mean monthly stream flows at this site based on the simulated daily stream flow are listed in Table 2. As shown, the estimated average annual discharge at the headworks is 12.9 m³/s, with monthly discharges varying from 4.6 m³/s in March to 35.7 m³/s in August.

Table 2: Estimated Long-Term Average Monthly Headworks Streamflow (m³/s)

Jan	Feb	Mar	Apr	May	Jun	Jul	Aug	Sep	Oct	Nov	Dec	Monthly Average
5.1	4.7	4.6	4.6	5.5	13.5	31.5	35.7	24.2	11.9	7.4	5.9	**12.9**

m³/s = cubic meter per second.
Source: Authors.

Estimation of Low Flow

Low-flow analysis was carried out to validate the dry season flow estimates from earlier design studies of the MRWDS. Low flows for different durations and return periods were estimated based on simulated time-series of daily discharges obtained from the SWAT model for the period 1982-2021 (Table 3). As per general practice and as recommended by the United States Environmental Protection Agency or setting permit discharge limits, the 7-day average low flow that occurs once every 10 years, i.e., 3.27 m³/s, was recommended as the low-flow value at the MRWDS headworks site.

Table 3: Estimated Low Flows (m³/s) at MRWDS Headworks Site

Return period (T), years	1 day	7 days	10 days	30 days
2	4.37	4.41	4.42	4.52
10	3.24	3.27	3.28	3.35
20	2.89	2.91	2.92	2.99

Source: Authors.

Estimation of Probable Maximum Precipitation

PMP is defined as the greatest depth of precipitation which is meteorologically possible for a given duration over a particular catchment area. For the Melamchi catchment, the PMP was estimated using a widely used statistical method developed by Hershfield[35] and recommended by the World Meteorological Organization (WMO).[36]

---

35   Hershfield, D.M. 1961. Estimating the Probable Maximum Precipitation. *Journal of Hydraulic Division: Proceedings of the American Society of Civil Engineers*, 87: 99-106.

36   WMO. 2009. *Manual on estimation of probable maximum precipitation (PMP)*. World Meteorological Organization (WMO). https://library.wmo.int/index.php?lvl=notice_display&id=1302.

For a 24-hour storm over the 161 km² catchment upstream of the MRWDS headworks site, a 1-day final PMP of 487 mm was estimated using daily precipitation data. Likewise, the 3-day, 10-day, and 1-month PMPs were estimated as 788 mm, 1,252 mm, and 1,660 mm, respectively. Return periods for extreme precipitation analyzed considering the available sparse regional statistical data yielded an estimated 10,000 years return period of 24-hour precipitation of 290 mm.

Estimation of Probable Maximum Flood

The PMF for the Melamchi catchment was estimated from the PMP. For this purpose, the daily rainfall data was disaggregated into hourly values to simulate the peak flood from the catchment, and the Johnson SB Distribution method was used to compute the accumulated rainfall for an hour of interest from a 24-hour precipitation data set. Feeding the SWAT model with a 1-day PMP distributed over 24-hour resulted in a PMF of 3,276 m³/s. The hydrograph of the PMF at Melamchi headworks site is depicted in Figure 19.

Figure 19: Probable Maximum Flood at Melamchi Headworks, Hourly Values

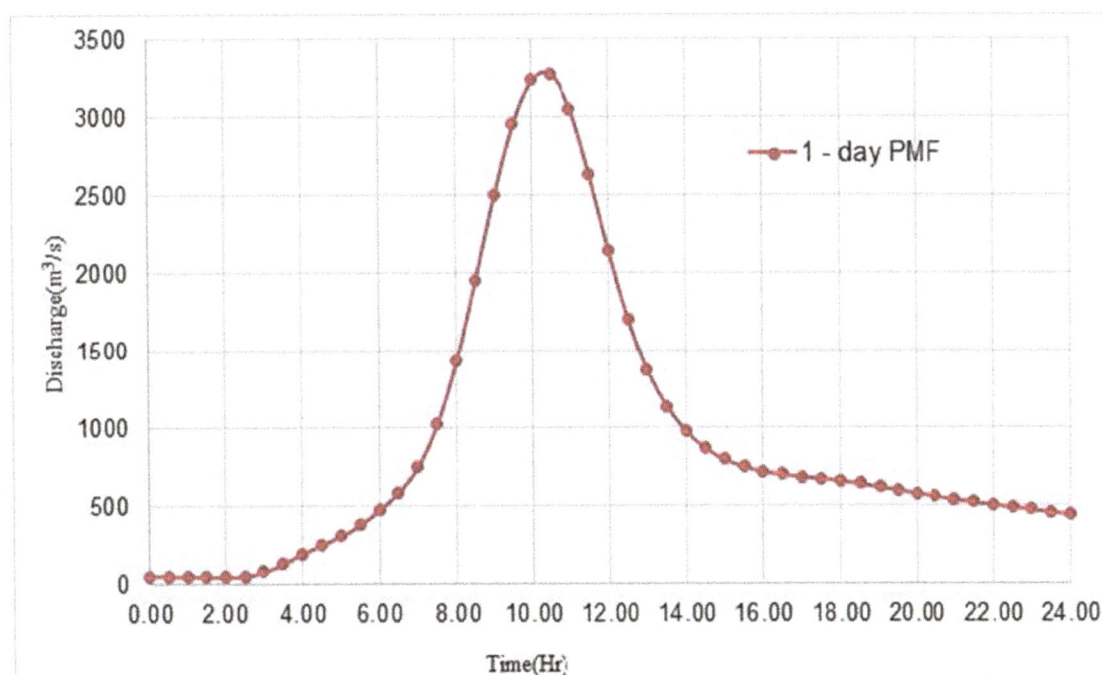

Hr = hour, m³/s = cubic meter per second, PMF = probable maximum flood.
Source: Authors.

To verify its estimated value, the PMF for the Melamchi catchment was compared with other river projects in Nepal based on PMFs per unit catchment area. The comparison showed that the estimated PMF for Melamchi follows the trend line for the other Nepalese projects.

Figure 20: Comparison of PMF between Projects in Nepal

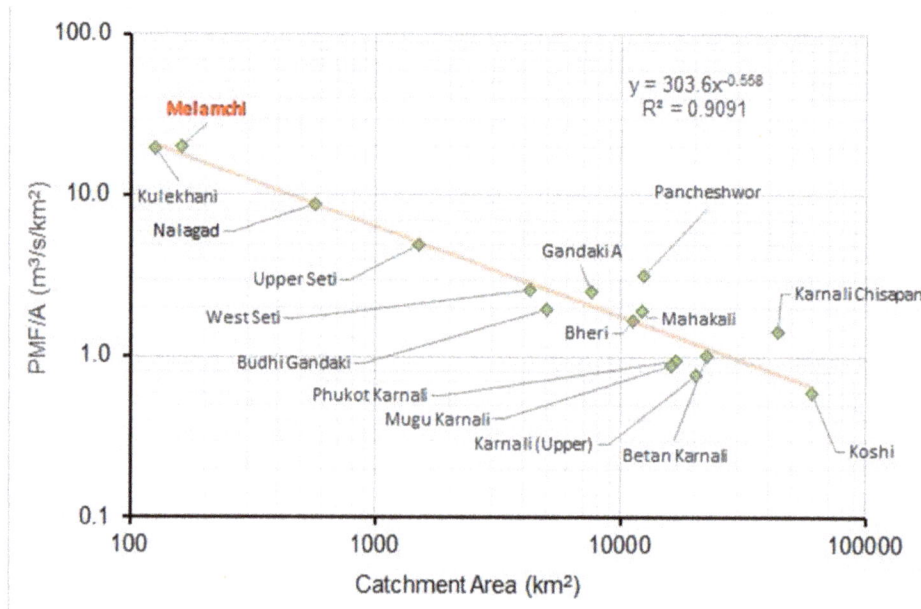

km² = square kilometer, m³/s/km² = cubic meter per second per square kilometer, PMF = probable maximum flood.
Note: The equation for the best fitted trendline, with vertical axis y and horizontal axis x, is depicted together with the regression value R² for how well the fitted line (light red) matches the observations (green dots).
Source: Authors.

Probable Maximum Outburst Flood

The PMF is calculated exclusively from consideration of precipitation. Investigations on the Melamchi events indicated, however, that the first flood event in 2021 was associated with both the rainfall-runoff process and a combination of GLOF, slope failure at Melamchi Ghyang, damming, and subsequent LDOF. Accordingly, a novel method for estimating the maximum discharge from such events was proposed in this study, and the resulting flood was referred to as probable maximum outburst flood (PMOF) (Figure 21).

In the proposed method, a two-stage approach was adopted for hydrological simulation of the first flood event. In the first stage, conventional rainfall-runoff processes were assumed using the SWAT model, and, in the second stage, additional GLOF and LDOF processes simulated with a simple spreadsheet model were integrated into SWAT as point discharge to enable it to reproduce the flood events.

For the LDOF simulations referred to above, the maximum probable water storage volume behind a landslide dam was calculated purely from geometric considerations of the V-shaped river valley, the height of neighboring ridges, and the geological characteristics over the slope. The resulting LDOF was calculated assuming the dam crest as a broad-crested weir. Using this approach, geological, geomorphological, and landslide information is incorporated in the flood estimation. Incorporation of such geological assessment along with hydrological assessments for maximum flood estimation is recommended for the design of new projects.

Figure 21: Proposed PMOF Approach for Estimating Maximum Discharge

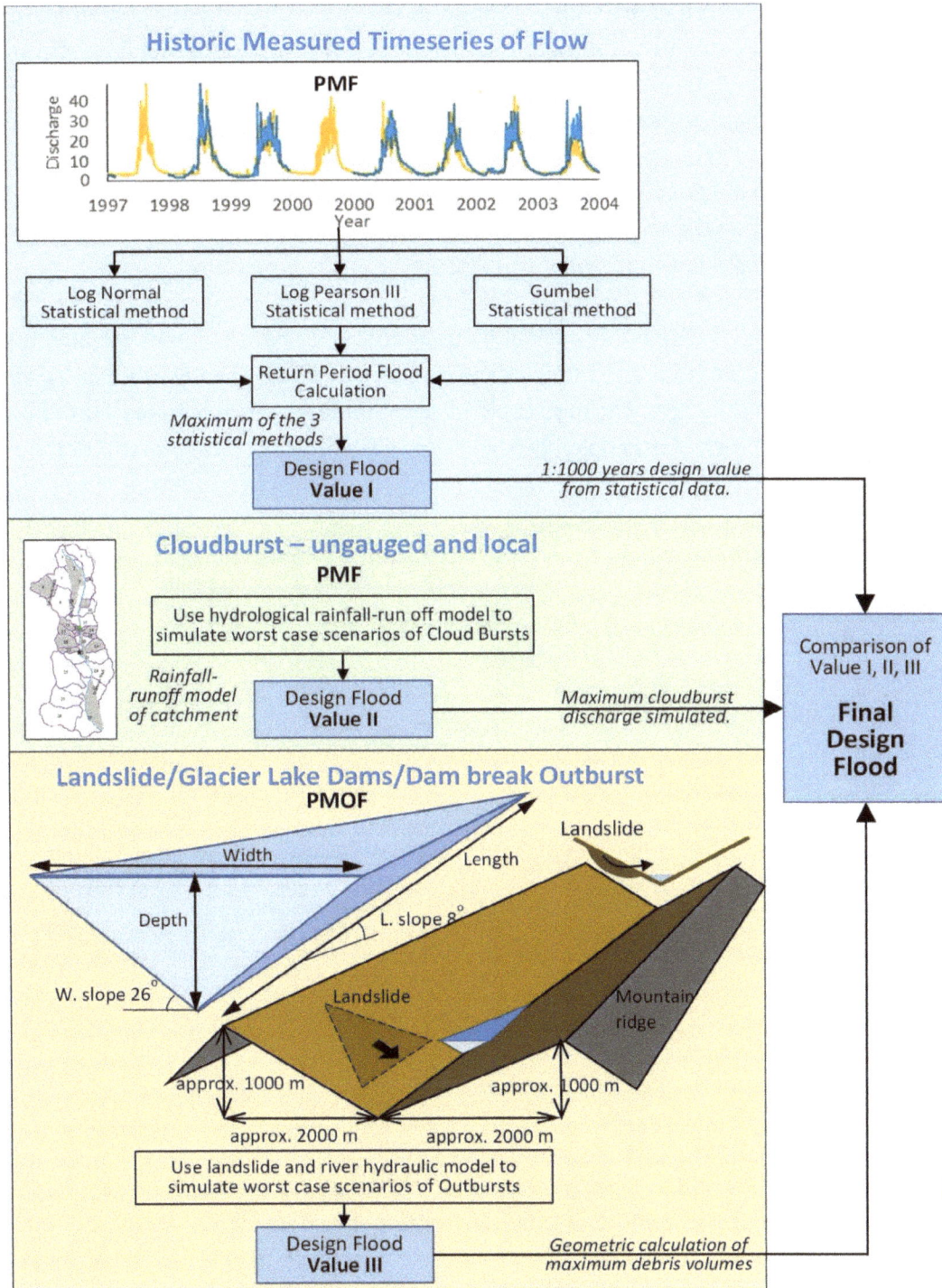

PMF = probable maximum flood, PMOF = probable maximum outburst flood.

m = meter

Source: Authors.

Hydrological Analysis of 2021 Flood Events

Analysis of rainfall data from June to August 2021, available at five meteorological stations in the Melamchi and nearby catchments, showed that rainfall in the area prior to the two events of 15 June and 1 August was not extreme. In the first case, continuous rainfall had occurred since 10 June 2021, creating a high degree of soil saturation. However, the maximum value of daily rainfall recorded in this period at all five meteorological stations was 77 mm, with the nearest station recording only 39 mm of rainfall on June 15. Likewise, in the second case, continuous rainfall had occurred for at least 2 weeks, but the maximum daily rainfall in the period at all five stations was 66 mm, recorded at the nearest station on 31 July 2021.

The records apparently indicate that rainfall by itself could not have created the large floods seen in 2021. Analysis of 5-minute rainfall data received from DHM in May 2022, however, showed that a peak discharge of 829 m³/s could have occurred for a few minutes. Furthermore, cloudburst-like phenomenon may possibly have occurred in the 44 km² catchment upstream of Melamchi Ghyang.

Figure 22: Rainfall Data from June to August 2021 in Melamchi Catchment

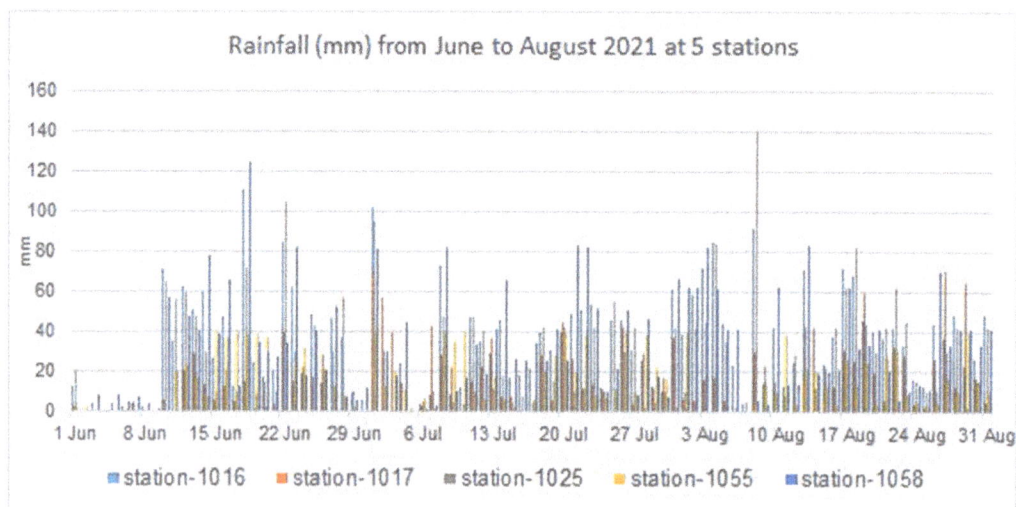

mm = millimeters.
Source: Authors.

For the hydrological analysis, the magnitudes of the 2021 flood events were estimated. For this purpose, a one-dimensional (1D) Hydrologic Engineering Center-River Analysis System (HEC-RAS) model was set up to develop discharge rating curves at two key locations near the headworks. After evaluation of the inundation extent simulated by HEC-RAS at different discharge levels and its qualitative validation from flood marks and local consultations, the magnitudes of both the 2021 flood events were estimated as **2,700 m³/s**.[37]

---

[37] Different views regarding the actual peak discharge in 2021 exist among reviewers of this report, which are acknowledged by the TA team. This estimate, which is based on the team's scientific analyses and available data, is however retained.

The estimated peak discharge was discussed with the NDRRMA/World Bank team, which had conducted a similar independent investigation along with flood modeling[38] using a two-dimensional (2D) HEC-RAS model to simulate the individual flood events in 2021. This team had also investigated the extent of flooded areas through simulations of depth, velocity, and delineated hazard zones, with the high viscosity dam breach mudflows simulated as non-Newtonian flow. The team had validated the results of its study through drone surveys and field observations and, in addition, had identified and mapped potentially dangerous areas using the model results. Not surprisingly, its analyses showed high sensitivity to the river bathymetry and floodplain topography, and to the percentage of debris in the river water.

Following scrutiny of the results from the two studies, the ADB and NDRRMA/World Bank teams agreed that the span of discharge values calculated by them in different ways illustrated the inherent uncertainties in their estimates. ADB's TA team, however, retained its estimate of 2,700 m³/s for the June 2021 flood. It is noted that the NDRRMA/World Bank team's modeling was important and useful in illustrating the inherent uncertainties as well as the observed flood patterns, including flow and flood depth characteristics. These observations were extensively used by ADB's TA team to optimize its model setup.

River Hydraulics and Sediment Transport Model Setup

Sediment Budget Analysis
A sediment budget analysis was performed to get insights into the trends of erosion, deposition, and sediment balance in the reach between Bhemathan and Melamchi Bazaar. The analysis was performed by computing the difference in ground and river Bed-Levels before and after the 2021 events. This computation was based on the 5 m resolution DEM created by Nepal's Department of Mines and Geology before the June 2021 event and the drone survey-based DEM prepared by NDRRMA after the 2021 flood events.

The DEM-based analysis showed an erosional trend in the reach upstream of the headworks up to Bhemathan and deposition in the reach downstream of the headworks up to Melamchi Bazaar. This trend is consistent with the observed field conditions. Noticeably, however, the analysis showed unrealistic local deposition and erosion depths that did not match the depths observed in the field. For instance, the analysis showed 25 m of deposition in the Bhemathan sedimentation zone as compared to around 5 m on average observed in the field. Similarly, in the headworks area, the analysis yielded around 40 m of deposition, whereas only about 22 m on average existed in this area. Therefore, quantification of sediment budget by comparing the DEMs proved unreliable, most likely due to inaccurate pre- and/or post-event data.

---

38   NDRRMA/World Bank. 2021. *Melamchi Flood Modeling*. Final report.

Figure 23: Erroneous Bed-Level Changes from DEM vs Improved Field Estimates

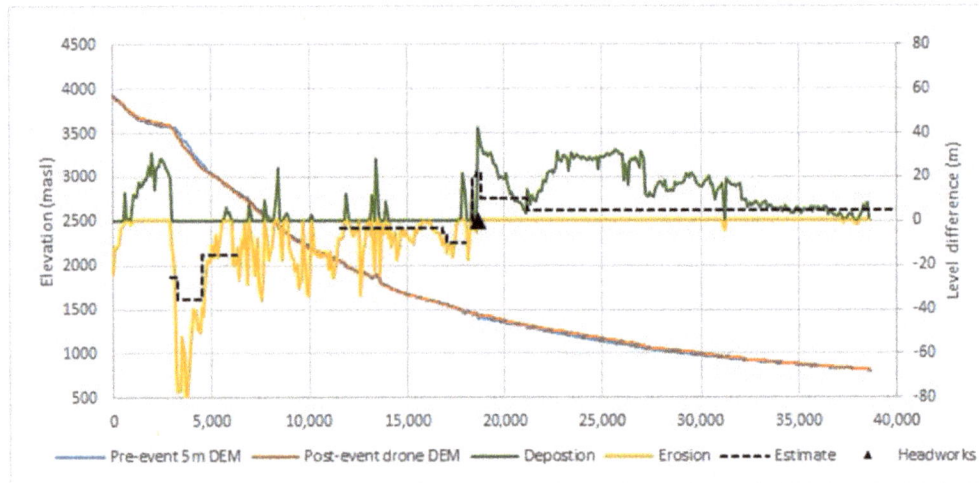

DEM = digital elevation model, m = meter, masl = meters above sea level.

Note: Black dashed line shows the field estimate.

Source: Authors.

In view of the above, erosion and deposition volumes were approximated by collating field data. For this purpose, the Bhemathan to Melamchi Bazaar reach was divided into several stretches. The surface width, and hence surface area, of the river channel, along with the average erosion and deposition depths, in each stretch were estimated through visual field observations and consultations with local people. Volume calculations based on these data showed that total erosion and deposition volumes between Bhemathan and Melamchi Bazaar were in the order of 20 million m³ and 17 million m³, respectively.

Table 4: Estimation of Sediment Budget, Bhemathan Knickpoint to Melamchi Bazaar

Location	Area m² (calculated from GIS)	Erosion		Deposition	
		Avg. bed change (m)	Erosion volume (m³)	Avg. bed change (m)	Deposition volume (m³)
Bhemathan amphitheater	31,400	-25	785,000	n/a	n/a
Old knickpoint - new knickpoint	174,819	-35	6,118,665	n/a	n/a
Downstream of old knickpoint (2.2 km reach)	726,978	-15	10,904,670	n/a	n/a
Narrow gorge -Melamchi Gyang	246,374	-3	739,122	n/a	n/a
Melamchi Gyang- Nakote	47,101	-3	141,303	n/a	n/a
Nakote	19,503	-5	97,515	n/a	n/a

Location	Area m² (calculated from GIS)	Erosion		Deposition	
		Avg. bed change (m)	Erosion volume (m³)	Avg. bed change (m)	Deposition volume (m³)
Nakote downstream – Sarkathali Bridge	87,860	-3	263,580	n/a	n/a
Sarkathali Bridge upstream	148,836	-5	744,180	n/a	n/a
Sarkathali Bridge downstream – narrow gorge headworks	45,069	-10	450,690	n/a	n/a
Headworks - upstream of weir - Narrow Gorge	13,139	n/a	n/a	20	262,780
Headworks (downstream of weir – 90-degree bend)	18,534	n/a	n/a	22	407,748
Headworks-Timbu Bazaar	202,591	n/a	n/a	10	2,025,910
Timbu Bazaar -Melamchi Bazaar	2,886,412	n/a	n/a	5	14,432,060
Total			20,244,725		17,128,498

GIS = geographic information system, m = meter, m³/s = cubic meter per second.

Source: Authors.

Although significant changes were evident locally, differences between the pre- and post-event river profiles were hardly visible on a regional scale. It was therefore concluded that the changes in the morphology of the river, including its average slope, width, bed sediment composition and riverbanks, after the 2021 flood events were insignificant on a catchment and river reach scale. Locally, however, large changes occurred in the riverbed elevation, width, bank erosion, sediment scour, and sediment deposit.

Choice of Model

The HEC-RAS numerical model system was used to analyze the hydraulics and sediment transport of Melamchi River. The model, which can simulate 1D, 2D, and combined 1D and 2D steady, quasi-unsteady, and unsteady flow conditions, integrates the Saint-Venant equations for the water flow and the Exner equation for bed updates.

Modeling of the Melamchi River is a challenging task owing to its supercritical flows caused by its steep gravel-bed and a series of hydraulic jumps. The river's high bed variability with riffles and pools further complicates its "modeling" was used in previous instances. Therefore, to simplify the modeling, a 1D quasi-unsteady simulation was performed to analyze the sediment transport of the Melamchi River. [39]

---

39   HEC-RAS, River Analysis System. *HEC-RAS River Analysis System User's Manual Version 6.0. February 2016.* https://www. hec.usace.army.mil/software/hec-ras/documentation.aspx.

Model Bathymetry

The 1D model was set up considering the 20 km reach between Bhemathan and the Melamchi camp located immediately downstream of the headworks. Its geometry file was generated by extracting the centerline of the main river channel and its cross-sections at 20 m intervals from the drone-based DEM developed in the NDRRMA/World Bank study just after the 2021 debris flood event. As the river discharge during the drone survey was not very high and water channels were very shallow, the DEM captured the topography well except along the main channel thalweg. The drone-based DEM revealed an average longitudinal slope of 12% in the Bhemathan-Melamchi Camp reach of the river, with around 20.3% slope in the Bhemathan-Nakote reach and around 6.8% in the Nakote-Melamchi Camp reach.

Figure 24: Model Domain for 1D Sediment Transport Modeling

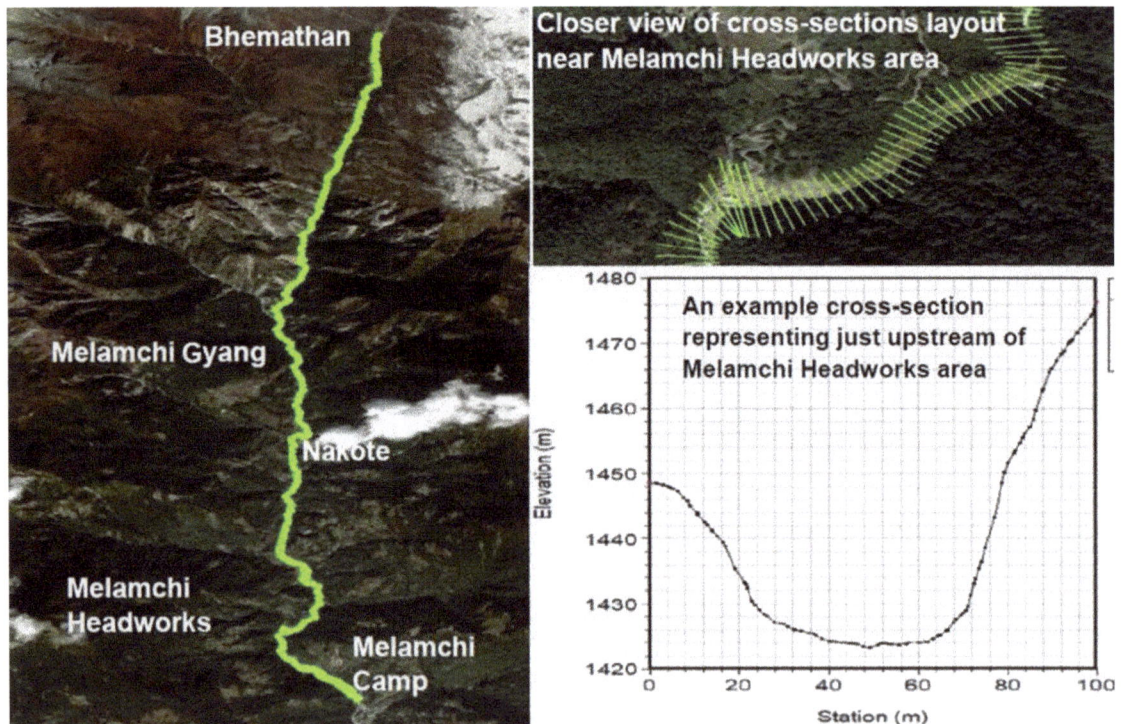

m = meter.
Source: Authors.

Figure 25: Longitudinal Profile of Key River Stretches

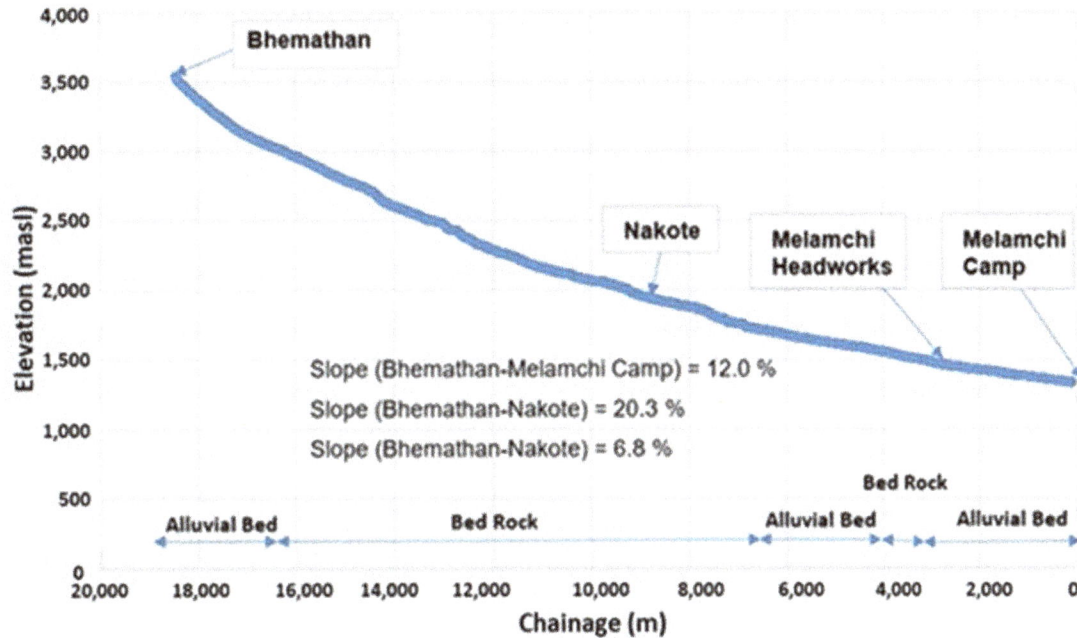

m = meter, masl = meters above sea level.
Note: Chainage refers to the distance in meters along the river from a defined reference point.

Source: Authors.

Flow Boundary

A time series of discharge reaching the observed peak of 2,700 m³/s was imposed at the upstream boundary (Bhemathan) in the HEC-RAS model. At the downstream boundary at Melamchi Camp, a discharge rating curve was specified assuming a water level slope equivalent to the bed slope at this location. This assumed the so-called natural depth.

Sediment Boundary

The sediment influx from the upstream boundary was defined as a sediment rating curve that relates the river's sediment transport rate with its discharge. This allowed the numerical model to dynamically select the sediment concentration value for each discharge and translate it into volumes of different classes of sediment using the PSD of the riverbed sediments measured at the Bhemathan deposit during the field investigations.

Computational Time step

The HEC-RAS model simulates hydrodynamics and updates the bed geometry at defined time intervals. It further subdivides computational intervals into bed mixing time steps. Considering the rapidly changing bed of the Melamchi River, the computational time step for the quasi-unsteady model was set to 0.02 hours (1.2 minutes) for discharges above 100 m³/s and to 0.10 hours (6 minutes) for discharges below 100 m³/s, and the mixing time step was set to 10 seconds.

Bed Roughness (Manning's Number)

A uniform Manning's roughness value of 0.05 s/m$^{1/3}$ was used for the main channel and banks. However, considering the uncertainties in Manning's roughness values and the sparse field data for calibrating the model, sensitivity analysis was performed with Manning's roughness value of 0.075.

Bed Sediment

As the particular interest area for the study was the MRWDS headworks, the PSD obtained from field measurements of bed sediment at the headworks deposits was used to specify the size of bed sediment for the entire model domain. The PSD curve was divided into 18 classes (clay to boulder) and included in the model. The 50% grain-size fraction (D50) and the 90% grain-size fraction (D90), representing the headworks area, in the model setup were 87 mm and 250 mm, respectively.

Sediment Transport Formula

The most commonly used Meyer-Peter Müller sediment transport formula for bed load transport analysis of gravel bed rivers was used to calculate the sediment transport. The Exner equation was used to determine sediment continuity at every cross-section and to update the riverbed elevation.

River Hydraulics and Sediment Transport Simulations

Simulated Erosion and Deposition

The numerical simulation result demonstrates a clear pattern toward erosion in the upstream catchment and deposition in the downstream, including the headworks area. It shows a sequence of alternating erosion and deposition in the narrow gorge, Sarkathali, and Nakote areas. However, the overall bed change pattern was mainly due to erosion, which is consistent with field observations. Noticeably, the model simulated deposition at Nakote, which may be due to the presence of a widened stretch of the river and narrow bottleneck just downstream of the Nakote bridge.

Figure 26: Longitudinal Simulated Bed-Level Change, Headworks to Nakote

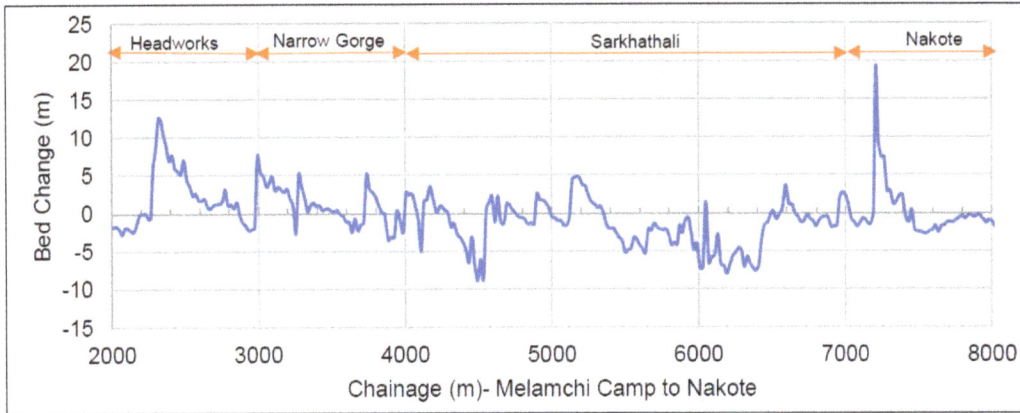

m = meter
Source: Authors.

Figure 27: Erosional and Depositional Trend Triggered by the 2021 Debris Flood Event

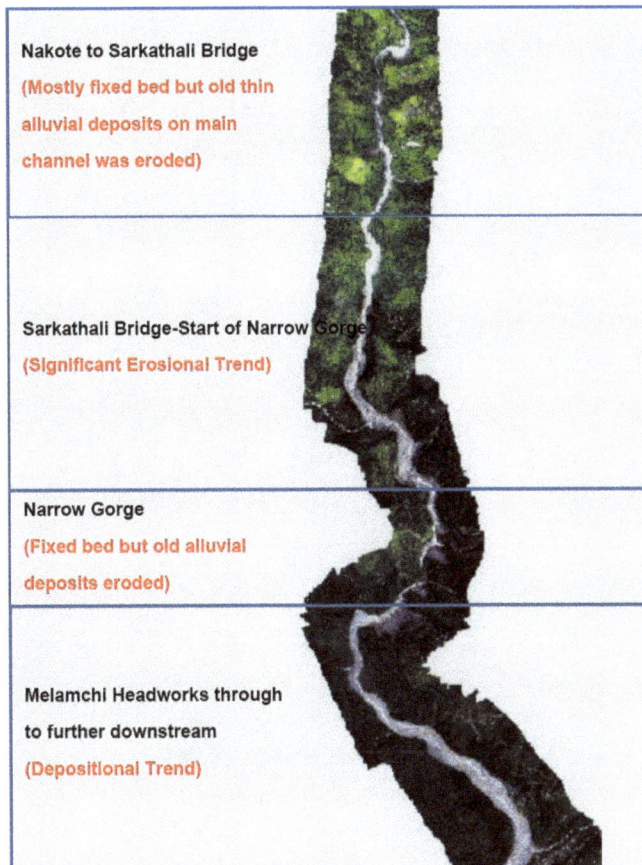

Source: Authors.

For the 300 m reach near the headworks (from its construction diversion tunnel to 300 m downstream), the model simulated channel deposition between 4 m and 10 m and overbank deposition between 4 m and to 15 m, whereas the observed maximum deposition in the headworks area during the 2021 event was around 22 m. In overall terms therefore, the numerical model was not able to accurately predict the maximum deposition observed in the river, even though contributions from both the June and July/August floods were simulated. However, the model simulated and captured the deposition channel and overbank trend well, consistent with field observations.

The observed deposition of 22 m within the 18,500 m² area in the vicinity of the headworks corresponds to a deposition volume of 400,000 m³. With the sediment balance calculations indicating another 16.4 million m³ deposited further downstream, the sediment entering the headworks area could have been around 17 million m³, of which less than 3% (coarsest material) would have deposited in the closest vicinity of the headworks area. The rest would have deposited further downstream (see Table 4).

Figure 28: Headworks Shown Partially Buried after June 2021 Flood

Source: Melamchi Water Supply Development Board, Government of Nepal.

Figure 29: Simulated Bed-Level Changes (Deposition) Near the Headworks

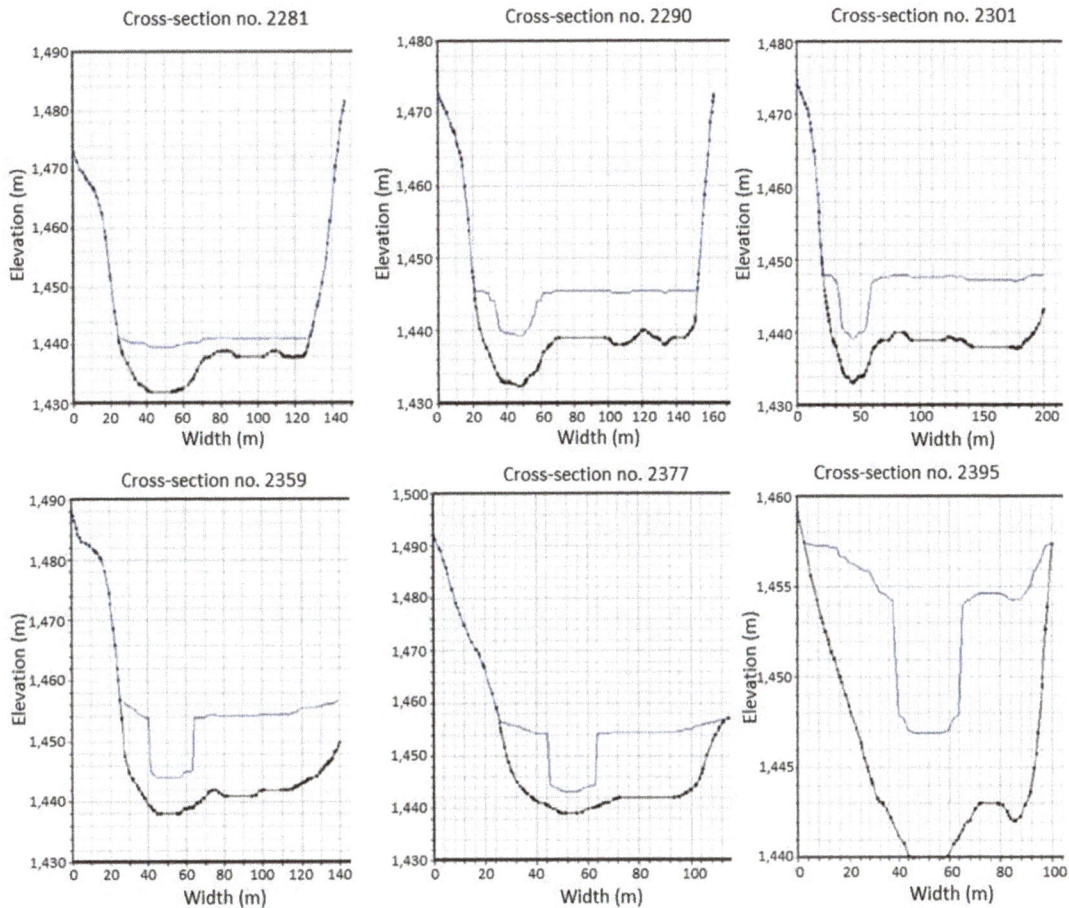

m = meter
Note: Black line shows the original cross-section, and blue line shows cross-section at the end of simulation.
Source: Authors.

Analysis of Different Influx Scenarios

Simulated Sediment Balance

A simulation assuming the same model parameter, i.e., n=0.05 s/m$^{1/3}$, and the Meyer–Peter Müller sediment transport formula was used to explore the response of Melamchi River to different influx scenarios. These scenarios ranged from 2,700 m^3/s down to 540 m^3/s and a flood discharge of less than 100 m^3/s (assuming no LDOF).

Figure 30: Simulated Discharge (Influx) Scenarios

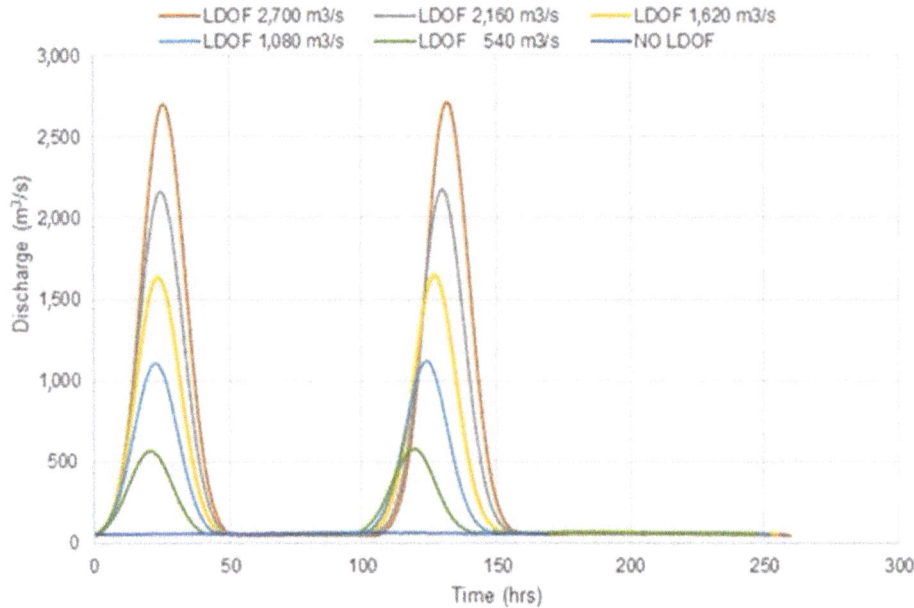

m³/s = cubic meter per second, LDOF = landslide dam outburst flood.
Source: Authors.

All simulated scenarios showed a similar pattern of deposition in the headworks area, the only difference between them being the deposition magnitude. The scenario without the LDOF resulted in insignificant sediment transport as compared to LDOF scenarios. The simulations therefore confirmed that the sudden flash floods induced by LDOFs or GLOFs resulting in debris floods were the main factors responsible for sediment transport in the river.

Figure 31: Headworks-Nakote Bed-Level Change with Different Discharge Scenarios

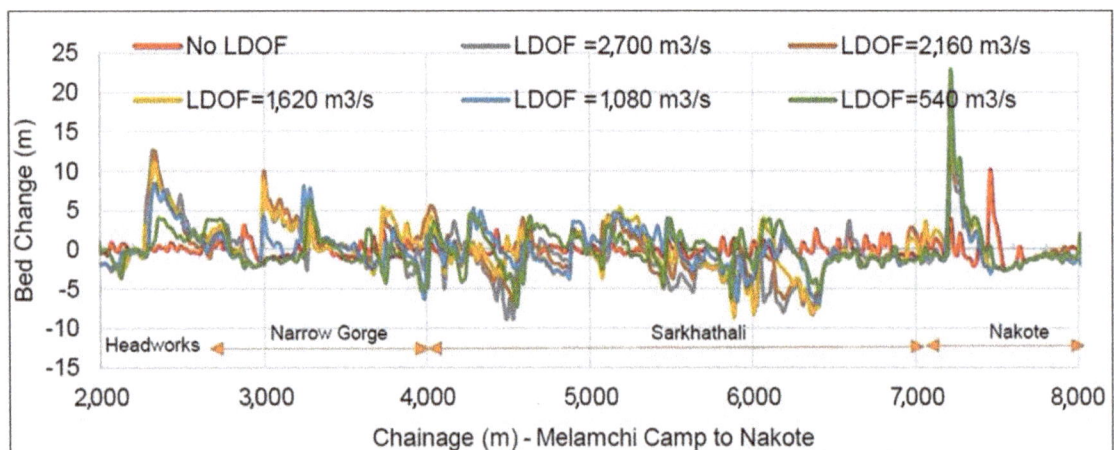

m = meter, LDOF = landslide dam outburst flood
Note: Chainage from 2,100 m to 2,800 m is the headworks area
Source: Authors.

Figure 32: Cumulation Sediment Transport with Different Influx Scenarios

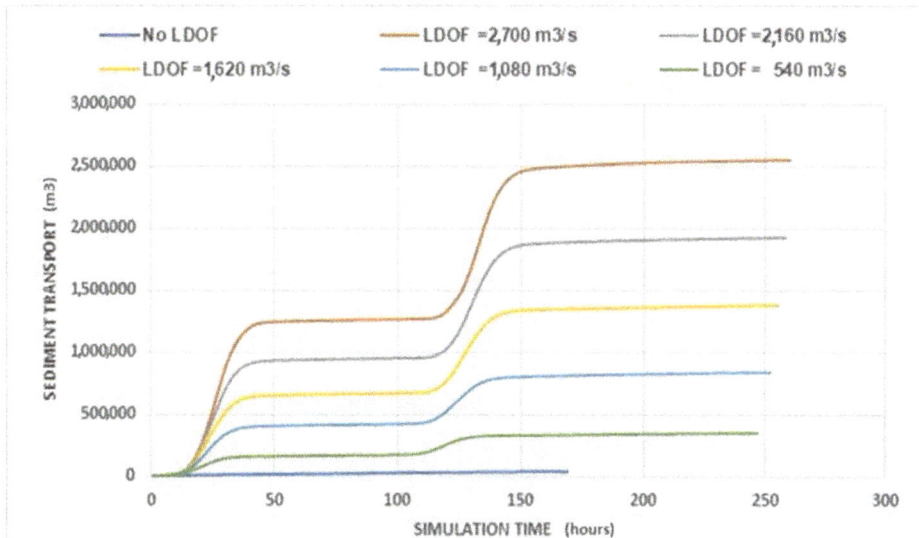

m³/s = cubic meter per second, LDOF = landslide dam outburst flood.
Source: Authors.

The simulated sediment transport of different fractions just upstream and downstream of the headworks is depicted in Table 5. The simulated gravel/cobble/boulder fraction accounts for 22% of the influx to the headworks area, sand for 23%, and silt and clay for 56%. The outflux is different, and the difference between influx and outflux shows the composition of deposited material. The gravel/cobble/boulder fraction shows the highest degree of deposition (74%) in the headworks area, evidently due to the sudden reduction in stream power when the river cross-section widens at the headworks. The percentage of silt and clay in the deposition is only 7%. As expected, most of the silt remained in suspension and was washed further downstream to the Melamchi catchment downstream of the headworks.

Table 5: Simulated Sediment Transport Upstream, Downstream of Headworks Area

Flow LDOF m3/s	Sediment In (CS3020)				Sediment Out (CS2254)				Sediment Balance (CS 2254–CS3020)			
	Total (m³)	Clay and Silt %	Sand %	Gravel/ Cobble/ Boulder %	Total (m³)	Clay and Silt %	Sand %	Gravel/ Cobble/ Boulder %	Total (m³)	Clay and Silt %	Sand %	Gravel/ Cobble/ Boulder %
2,700	2,584,640	55	23	22	2,326,666	61	23	16	257,974	7	19	74
2,160	1,930,860	54	23	23	1,747,290	59	23	18	183,571	8	20	72
1,620	1,389,349	52	22	26	1,194,559	59	23	18	194,789	7	18	75
1,080	882,300	48	22	30	742,141	55	23	22	140,159	6	17	77
540 m	454,269	38	22	40	382,584	44	23	33	71,685	7	17	76
No LDOF	45,436	32	31	27	35,908	39	34	27	9,527	7	20	73

CS3020 = cross-section at inflow to headworks area, CS2254 = cross-section at outflow from headworks area
m³/s = cubic meter per second, LDOF = landslide dam outburst flood
Source: Authors.

Flood Travel Time and Velocity

A quasi-unsteady model does not simulate the flood travel time, for instance, of a LDOF or GLOF event. Therefore, a fully dynamic 2D HEC-RAS model was set up covering the area from Bhemathan to the headworks to the Melamchi camp near Timbu Bazaar to simulate the flood time assuming only clear water. Simulations of the June 2015 event showed that a flood wave from Bhemathan would take 7 minutes to reach Nakote, 12 minutes to reach the headworks, and 17 minutes to reach Timbu Bazaar. The corresponding average travel velocity of the 2,700 m³/s flood discharge was 25 m/s for the Bhemathan-Nakote reach, 22 m/s for the Bhemathan-Melamchi headworks reach, and 18.1 m/s for the Bhemathan-Timbu Bazaar reach.

Figure 33: Flood Wave at 2,700 m³/s - Narrow Gorge to Downstream of Headworks

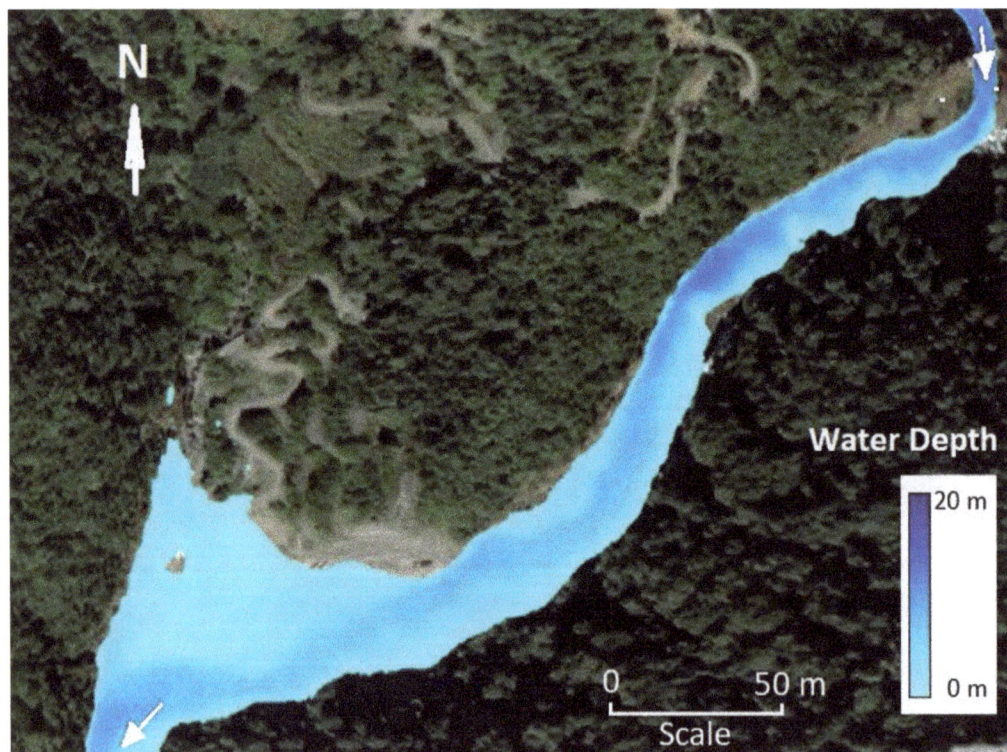

m = meter.
Source: Authors.

5. ESTABLISHMENT OF NEW DESIGN BASIS

Multi-Hazard Risk Assessment

From earthquakes to floods and landslides, the Melamchi River is exposed to multiple hazards that could potentially domino into cascading events and threaten the water it supplies to the Kathmandu Valey. Considering the mounting impacts of climate change risks exacerbating these hazards, vulnerable sites and locations need to be identified so that mitigation measures can be taken to better prepare them.

To understand how hazards in the area combine and interact, the TA team used event tree analysis to examine cause-and-effect relationships in cascading events. This analysis provides an understanding of event-triggering factors, considers possible time delays between subsequent events, and can be used to outline possible interventions to break event chains before they spiral into major, multi-hazard disasters. It considers the likelihood of individual hazards occurring and can be used to map the combined susceptibility of multi-hazard events to enable geographical identification of vulnerable sites.

Event Tree Analysis of Multi-Hazard Cause and Effect

An example of an event tree analysis considering eight scenarios triggered by a sequence of events is illustrated in the schematic in Figure 34. For each trigger event, there is a probability of occurrence, and, for a chain of events, there is a probability for the next event to happen.

Figure 34: Event Tree Analysis of Potential Melamchi River Basin Hazard Scenarios

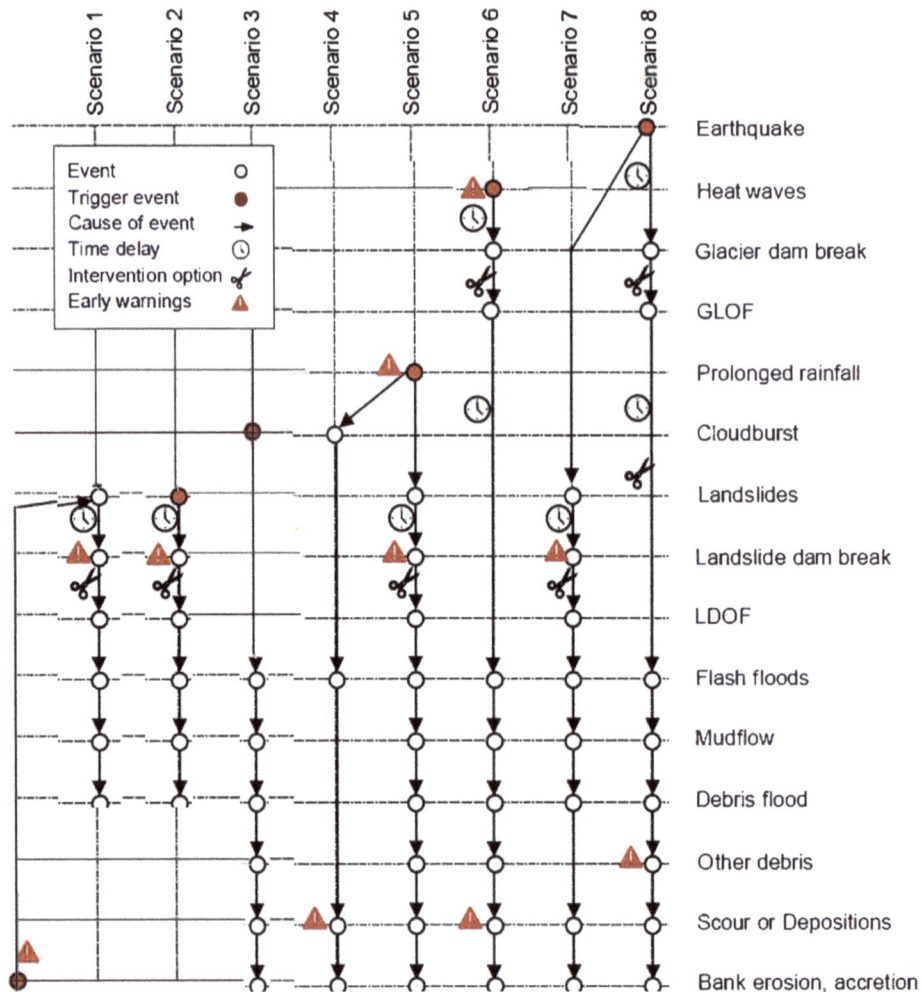

GLOF = glacial lake outburst flood, LDOF = landslide dam outburst flood.
Source: Authors.

The schematic illustrates, for instance, that earthquakes and temperature rise can trigger events like avalanches, landslides, moraine dam breaks, and debris flows that could displace water in glacial lakes. In another scenario, prolonged rainfall could trigger a landslide leading to formation of a natural dam.

In the first case, the schematic illustrates that the time lapse between events, such as a sudden temperature rise and subsequent ice melt, could be used by authorities to prevent knock-on effects such as GLOFs.

In the second case, the time required for a lake to build up behind the dam could be utilized to monitor lake levels. Suitable interventions such as draining lakes, dredging deposits, and blasting natural dams could be undertaken before the dam breaks. Likewise, the occurrence of some events could be monitored with an early warning system (EWS) that, if properly implemented, could break the cascading hazard chain by evacuating downstream areas to protect lives, property, and infrastructure.

It should be noted, however, that determination of the likelihood or probability of trigger events such as earthquakes is highly uncertain, and this uncertainty is further amplified if cascading events are considered. Likewise, the lack of historic time series of landslides from past decades impedes development of return period statistics, which therefore requires the use of relative susceptibility categories such as VHS (very high susceptibility), HS (high), MS (moderate), LS (low), and VLS (very low susceptibility). This is true for the Melamchi and adjacent catchments for which accurate records of historic events do not exist.

Decision Tree Analysis for Multi-Hazard Assessment

Given the nature of events seen in the Melamchi catchment, a multi-hazard assessment framework, rather than consideration of single events in isolation, is essential for obtaining a comprehensive picture of hazards and risks. Accordingly, an overlay analysis of different disaster susceptibility scenarios was performed to develop a zonation map using a decision tree tool.

For the overlay analysis, susceptibility zones were defined based on field and remote sensing assessments to characterize the contributions to the cascading hazards during floods in the Melamchi River catchment. In addition, five categories of susceptibility quantifications, ranging from very low, low, moderate, high, and very high, were defined based on a decision tree tool. Thereafter, all the classified geographic information system (GIS) map layers were overlaid, and the summed values were again classified into five categories of multi-hazard scenarios. The final overlaid multi-hazard map considering the multi-hazard scenario is shown in Figure 43.

Figure 35: Decision Tree Analysis of Melamchi Catchment Disaster Susceptibility

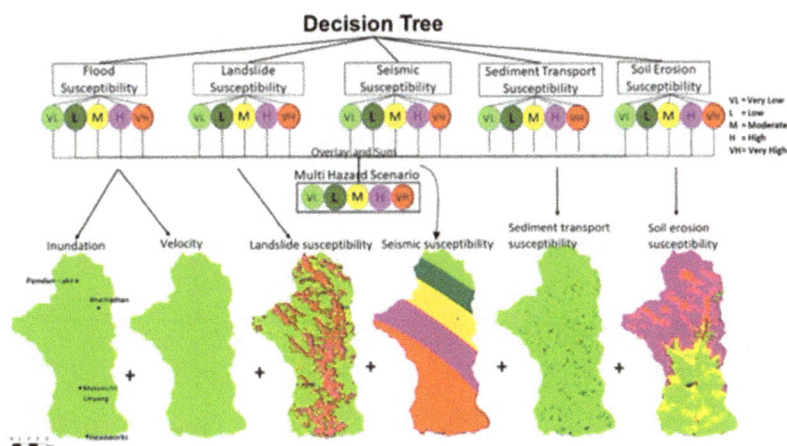

H = high susceptibility, L = low susceptibility, M = moderate susceptibility, VH = very high susceptibility, VL = very low susceptibility.
Source: Authors.

Figure 36: Longitudinal Profile of Predicted Melamchi Multi-Hazard Index

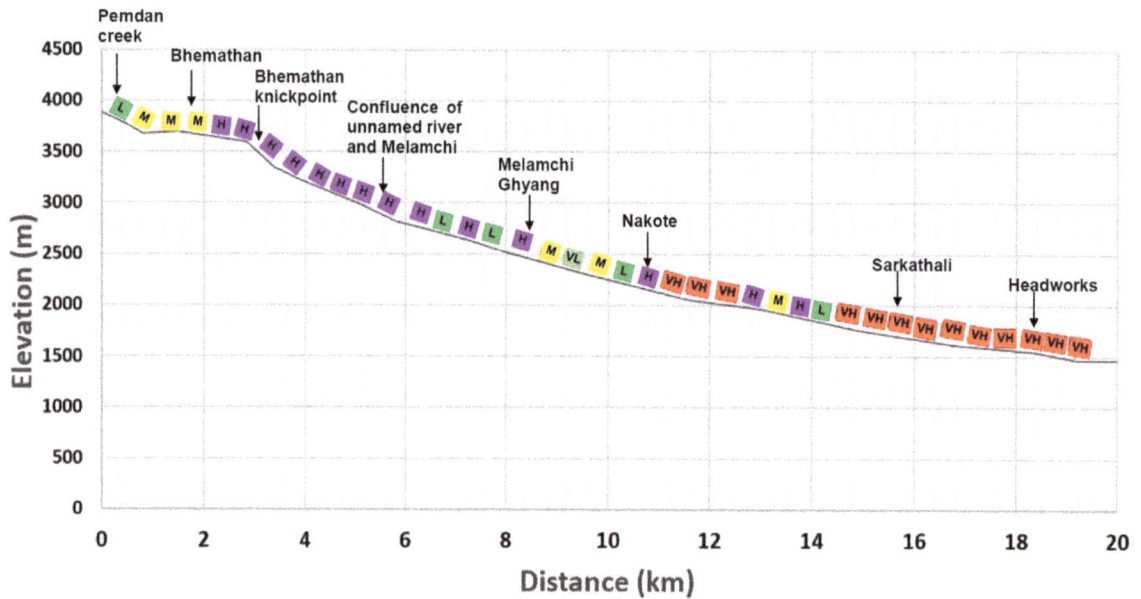

m = meter, km = kilometers.
Note: Bright green = very low susceptibility, dark green = low susceptibility, yellow = moderate susceptibility, pink = high susceptibility, red = very high.
Source: Authors.

Not surprisingly, the multi-hazard scenario analysis reconfirmed that the area from Bhemathan to the headworks had a very high multi-hazard value due to floods and bank erosion risks. Similarly, the Pemdan and Namsan creeks were classified as moderate-to-high multi-hazard zones.

Based on the event and decision tree analyses, five sites were highlighted as vulnerable locations that require specific attention:

- The headworks left bank landslide
- The Bhemathan area
- The Melamchi Gyang landslide
- Confluence points of Melamchi River with its tributaries upstream of the headworks
- Tributaries in the uppermost part of the catchment, upstream of Bhemathan

Figure 37: Catchment Multi-Hazard Zonation Map Using Decision Tree Analysis

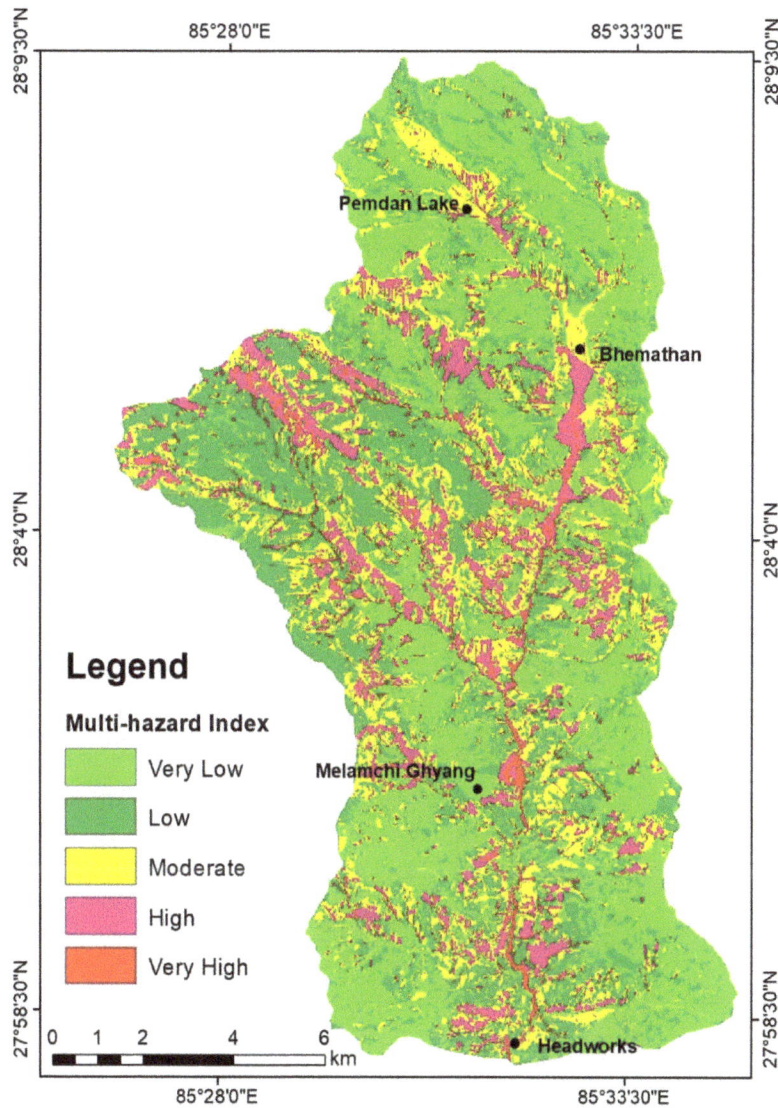

Source: Authors.

Results of Multi-Hazard Risk Assessment

Risk Identification

The events of 2021 were triggered by various hazards that interacted and cascaded to cause significant damage. These are some of the hazards the Melamchi catchment now faces:

- Earthquakes
- Cloudbursts
- Flooding

- Prolonged rainfall
- Snowmelt
- Landslides and landslide damming
- Debris and mudflows
- Heatwaves, causing impact snowmelt runoff during dry season
- GLOF
- LDOF
- Channel scouring and channel siltation
- Bank erosion and bank accretion
- Drought
- Vegetation and organic debris clogging infrastructure
- Infrastructure defects due to inadequate design
- Water contamination
- Vandalism and "artificial" environmental, technological, biological hazards

In conjunction with the type of hazard, the intensity of each hazard and its characteristics, such as duration, time of year, cascading effects etc., need to be considered to reduce the impact of the hazards on the elements at risk, including critical infrastructure and communities. In the case of the MWSP, the element at risk is the headworks whose vulnerability is defined by its design criteria. Unfortunately, the design criteria originally adopted for the headworks did not adequately address the conditions arising during the 2021 monsoon, resulting in extensive damage. Therefore, the headworks design criteria for its rehabilitation or repurposing should focus on conditions of extreme flood scenarios with consideration of maximum water levels, flow velocities, turbidity, and depths of sediment deposition and scour in the headworks structures.

Another core design criterion for the headworks relate to extreme drought scenarios, including minimum water levels, minimum river flow, and diurnal fluctuations of flow during low flows. The type and level of water contaminants and the amounts of debris that can potentially clog the intake structures also need to be factored in.

Vulnerability to hazards grows with increasing hazard intensity, which means vulnerability can be reduced if the hazard intensity is mitigated. For critical infrastructures such as the Melamchi headworks, reduction in vulnerability can be attained through three strategies- first, by reducing hazard intensity through catchment treatment; second, by reducing hazard vulnerability through "stronger" or resilient design; and third, by reducing hazard intensity and vulnerability through a combination of the first and second strategies.

The first strategy will reduce the catchment susceptibility to hazards and reduce downstream risks to life and property, but will only slightly reduce the vulnerability of a "weak" structure. The second strategy will leave hazards and downstream risks intact but will significantly reduce the impacts of hazards on the structure. The third strategy will yield the benefits of both the first and second strategies.

Figure 38: Vulnerability of Element-at-Risk as a Function of Hazard Intensity

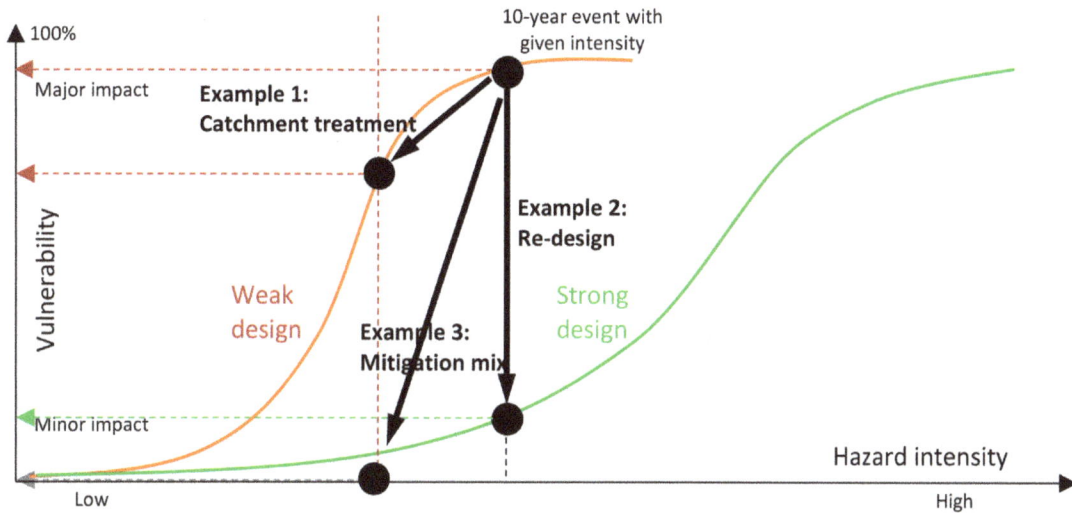

Source: Authors.

Risk Evaluation

Due to the inherent uncertainties in its determination, the likelihood or probability of occurrence of each cascading chain of events in the Melamchi catchment was not quantified numerically in terms of absolute probabilities. Instead, the extreme impact of these events was considered through a risk matrix with various categories of hazard frequencies and hazard impacts.

Figure 39: Risk Matrix Illustrating Hazard Frequency and Impacts

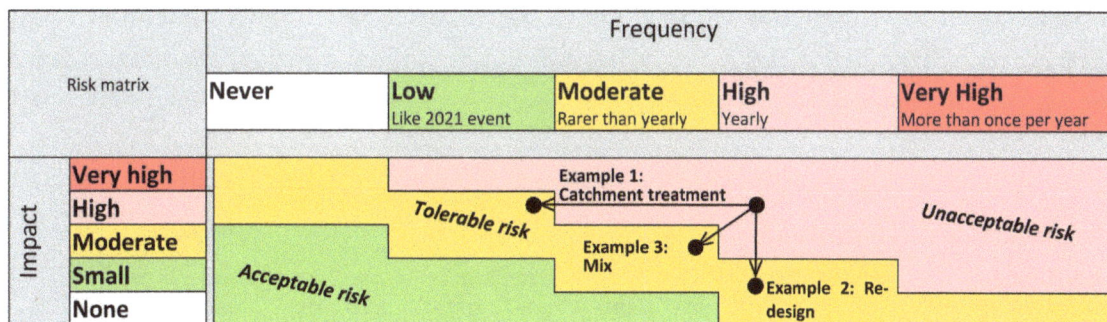

Note: Dots and arrows show mitigation measures.
Source: Authors.

The frequency criteria in the risk matrix ranged from low (similar to the 2021 event), moderate (less than yearly), high (yearly), and very high (more than once per year). These criteria were judged based on the model simulations (although sufficient statistical data was not available) and on observed events in 2021. Likewise, the hazard impact criteria were defined considering all relevant stakeholders through the cost of days without water production, cost of repair work like dredging, embankment reconstruction, structural repairs, cost of lost or damaged equipment, and the cost of casualties.

During possible design modifications after the detailed damage assessment of the headworks structures, risk evaluation using the hazard frequency versus impact matrix will consider the following design parameters related to the headworks:

1. High water levels during floods
2. Low water levels during drought (e.g., for establishing the intake sill level)
3. High flow velocities
4. Low river flows (drought) to determine maximum abstraction
5. Elevated turbidity levels of water
6. Sediment deposition height in front of structures (debris floods and extreme floods)
7. Scour near structures (extreme floods)

The design parameter analyses will be factored into the preliminary headworks designs in phase 2 and in any detailed redesign where a performance-based approach is adopted to manage the varying degrees of impacts from a wide range of loading conditions.

Resilience through Combinations of Risk Mitigation Measures

The risk assessment shows that a combination of structural and non-structural measures is needed in the Melamchi River catchment, the river channel, and at the headworks, to increase the resilience of critical MRWDS structures and protect them from future events. Such measures include improved monitoring of potential hazards, stronger management systems to ensure a hazard does not spark a disaster, and engineering solutions to minimize the impact of events. Apparently, the damage to the MRWDS headworks from the 2021 events resulted from the exclusive focus of its original design on the headworks area without sufficient orientation toward the catchment and the river.

Table 6: Combinations of Mitigation Measures to Be Further Investigated

Short/Mid/Long Term	Catchment	River	Headworks
Monitoring Early warning	Meteorological stations, radars, CCTV, drones, Earth Observation, inclinometers	Hydrological stations, real-time level meters, *silt-ation/scour probe monitoring suspended sediment	Flow meters, pressure and level meters, vibration sensors, turbidity meter
Management Swift response to events	Siphon dewatering of lake, slope scaling, landslide management, catch drain cleaning, dam breach initiation	Flood management, sediment management, dredging, excavation, repair of river works	Control automation, temporary shutdowns, sediment bypassing, desander cleaning, debris removal*

Short/Mid/Long Term	Catchment	River	Headworks
Stabilization Resilient design	Inter-basin water transfer, land slope stabilization, slope drainage,* land use regulation	River training works, revetments, groynes dams, check dams, embankments	Intake layout, desander design gate controls, weirs, flood walls

Note: * Refers to urgent actions recommended to MWSDB

Source: Authors

The combination of mitigation measures has a third dimension, timescale. While sustainable solutions to the identified hazards should be its ultimate goal, MWSDB has to consider short- and medium-term solutions, in addition to those for the long term, to meet the pressing need to supply water to the Kathmandu Valley. Accordingly, urgent short-term measures indicated in the table have been elaborated upon and communicated to MWSDB for its consideration. Key proposals for the medium and long term, which will be considered in phase 2, include

- Improved hydraulic design basis considering PMOF,
- Monitoring and multi-hazard EWS,
- Preparedness and response plans, and
- Community-based disaster risk management (DRM).

An improved hydraulic design basis should consider a more conservative possible maximum discharge by expanding the conventional precipitation-based PMF to PMOF. This is based on multi-hazard assessment of GLOFs and LDOFs in conjunction with precipitation analyses that also consider moraine depositions and their failure or disintegration due to heavy rainfall or snowmelt. Neglecting GLOF and LDOF will severely underestimate the maximum discharge. In addition, changes in the river morphology in terms of riverbed scour and deposition, as well as bank erosion and accretion, should be included in the hydraulic design basis.

Monitoring and early warning of multi-hazards in a disaster-prone area like the Melamchi River catchment requires a high-density network of monitoring stations. Monitoring should include ground stations for hydrological and meteorological data as well as modern technology such as Earth Observation for land displacement data. Systems should also be able to cope with power supply fluctuations and may be combined with community-based systems via mobile phone. Critical data on water levels in catchment lakes, river flows, or sudden landslides must be in real time or near real time.

Preparedness and Response Plans are essential if the monitoring data and early warnings are to be meaningfully used. For a range of multi-hazard combinations, there should be a set of contingency plans which could be initiated with data signal triggers. This could include closing intake gates in the event of a warning about an approaching high sediment-laden flood or mudflow event.

Community-based DRM has proven to be very efficient for managing risks in remote localities. Local people and workers on site can quickly communicate information by mobile phone and switch vital equipment on or off. Dubbed the "Internet of people," this can be more effective than digital technologies such as the Internet of Things which are vulnerable to fluctuations in power supply and mobile coverage and may be unreliable at high altitudes. By creating local engagement and ownership, community-based DRM can help protect assets at remote sites. It could also help reduce the impact of events on downstream structures and thereby cut the cost and time of rehabilitation.

New Criteria and Approach

Based on conventional hydrological studies conducted in the early 2000s, a design flood of 600 m^3/s, corresponding to a 10,000-year return period, was adopted for the Melamchi headworks. This flood frequency was selected considering the importance of the headworks for water supply in Kathmandu, and the corresponding flood discharge was estimated based on conventional considerations in practice when the project was designed. However, this headworks design discharge was exceeded twice in 2021- by 4.5 times if the ADB TA team's estimate is considered, and by almost seven times considering the estimate of the NDRRMA/World Bank team. As such, reasons for exceedance of the design flood need to be understood and a realistic approach for estimating design floods developed.

As previously discussed, conventional methods for estimating flood discharge are based purely on hydrological and meteorological considerations, with the PMF calculated exclusively from considerations of precipitation. These methods yield satisfactory results if they are based on adequate historical data that capture short duration and high-intensity rainfall events (e.g., cloudbursts) and physical processes associated with outbursts (e.g., GLOF and LDOF). In the case of Himalayan catchments, such data is generally not available, and therefore the effect of outburst events are not reflected in hydrological estimates. This leads to lower estimates of flood discharges, and consequently, increased risks from multiple hazards.

The 2021 flood events in the Melamchi catchment clearly indicate that the maximum flood estimation for such catchments should include various processes in geology, geomorphology, hydrology (including cloudbursts), glacial lake dam failure, etc, that influence multi-hazards but are not adequately captured in historical statistics of the catchment. Accordingly, the PMOF concept introduced in this study (Figure 21) utilizes the following approach to incorporate geological and geomorphological assessments with hydrological assessments for maximum flood estimation:

Estimation of LDOF: The maximum possible storage volume of water behind each potential landslide-induced dam across the river is estimated from geometric considerations of possible dimensions of the dam in the river valley. These dimensions are estimated considering the availability of landslide material on adjacent slopes through geological and geomorphological assessment of the overburden thickness above the bedrock on the slopes. Finally, the LDOF is simulated from a dam break model or simply using a broad-crested weir formula. Outburst floods can also be GLOFs caused by the failure of a dam containing a glacial lake.

Estimation of GLOF: The magnitude of GLOFs can be estimated using the LDOF estimation procedure considering the outburst to be triggered by the failure of a dam containing a glacial lake.

Estimation of cloudburst floods: The magnitude of flash floods resulting from cloudbursts are calculated using hydrological modeling.

Estimation of PMF: For comparison, the conventional PMF is calculated from rainfall-runoff statistics only. It should be noted that hydrological data of sufficient quality covering a long period will inherently contain LDOF, GLOF, and cloudburst flashfloods, thereby possibly foregoing the need for their assessment as described above. In Nepal, however, such data are generally not available.

Selection of maximum flood discharge: The maximum flood discharge for the given catchment is selected based on a rational combination and comparison of the various floods.

The PMOF approach was applied to the Melamchi catchment to estimate its maximum flood. The methodology and results of the application are discussed in the following sections.

Landslide Dam Lakes and Glacial Lake Outburst Floods

Volume of Landslide Dam Lakes

Results of the analysis of the sites and characteristics of landslides and slopes in the Melamchi catchment (Section 3.3) was used to calculate the possibility of landslide dams that could block the river and result in the development of ponds or reservoirs. These calculations involved estimations of landslide material available for creating dams, the height of such dams assuming a certain geometry, and the storage volume generated upstream. The actual runout of each landslide was not considered. Instead, it was assumed, as a conservative estimate, that all landslide material would be available to form a "perfect" dam.

Future Topography of Bhemathan

The topography of Bhemathan is subject to changes due to both sedimentation and erosion processes that may not be synchronized, with some years experiencing more erosion than deposition and vice versa. Topography data indicates that the lower 800 m of the Bhemathan plain was relatively flat before 2021 but now has a slope of 5%. Calculations show that a dam of 15 m height will create a storage volume of 616,000 m³ with the current slope, whereas the historic slope would have resulted in a higher volume.

Glacial Lake Volumes

The existing glacial lakes in the Melamchi River catchment are small. With an estimated area of 14,700 m², even the Pemdan Creek lake, which was drained in June 2021, would have had a storage volume of only 30,000 m³ to 45,000 m³ for storage depths of 2 to 3 m. However, existing glaciers in the catchment are large and may potentially create larger glacial lakes. For instance, the glacier in the upper reaches of the Melamchi River has an estimated area of 1.3 million m². If 800,000 square meters of this area convert into a glacial lake of 0.5 m depth on average, the potential storage volume of a future glacial lake at this location would be 400,000 m³.

Figure 40: Glacier at the Upper Reach of the Melamchi River

Source: Authors, using background image from Google Earth.

Calculations of LDOF and GLOF

Following estimation of the storage volumes of potential landslide dams and glacial lakes, dam break simulations were conducted. These assumed that a cascade of dams would breach simultaneously and generate the most extreme, although exceedingly rare, scenario. The simulations assumed various dam break failure modes such as overtopping or piping, where water seeps through the dam wall.

The speed of earthen dam erosion depends on various parameters whose estimation poses substantial uncertainties. However, the hydrographs of these events are similar, with a steady increase to a certain outburst peak discharge, and thereafter a decrease until the reservoir is emptied. A simplification in the simulation is, therefore, to assume a triangular hydrograph and calculate the peak discharge Q_{peak} from the storage volume $V_{storage}$ and time for breaching ΔT_{breach} using the following:

$$Q_{peak} = \frac{V_{storage}}{\frac{1}{2}\Delta T_{breach}}.$$

The results for the Melamchi catchment using the above relation are presented in Table 7.

Table 7: Reservoir Storage Volume with Dam Break Outburst Flood Calculation

Number	Reservoir ID	Dam Height	Storage Volume	Outburst Flood (Breach 30 minutes)	Outburst Flood (Breach 60 minutes)
		m	m³	m³/s	m³/s
1	LS20 at Bhemathan (present topography)	15	616,767	684	342
2	LS2 (upstream of headworks)	25	248,700	276	138
3	LS1 (near headworks)	25	241,748	268	134
4	LS4 (downstream Sarkathali)	20	150,496	166	83
5	LS14.1 (downstream Nakote)	25	148,613	165	83
6	LS20 (future topography of Bhemathan)	5	1,600,000	1,778	889
7	Glacier (future lake upstream)	0.5	400,000	444	222
Sum (2–7)	Cascade of LDOF and GLOF breaching with simultaneity		2,798,557	3,097	1,549
Sum (6–7)	Only Bhemathan and upstream glacial lakes		2,000,000	2,222	1,111

GLOF = glacial lake outburst flood, LDOF = landslide dam outburst flood, m³ = cubic meter, m³/s = cubic meter per second.

Source: Authors.

Table 7 indicates that the outburst outflow from Bhemathan is key to defining the maximum discharge at the headworks area. As such, the simulated accumulated peak discharge at the headworks is shown as a function of the storage volume at Bhemathan and upstream in Figure 41.

Figure 41: Discharge from Simulated Bhemathan Lake Outburst

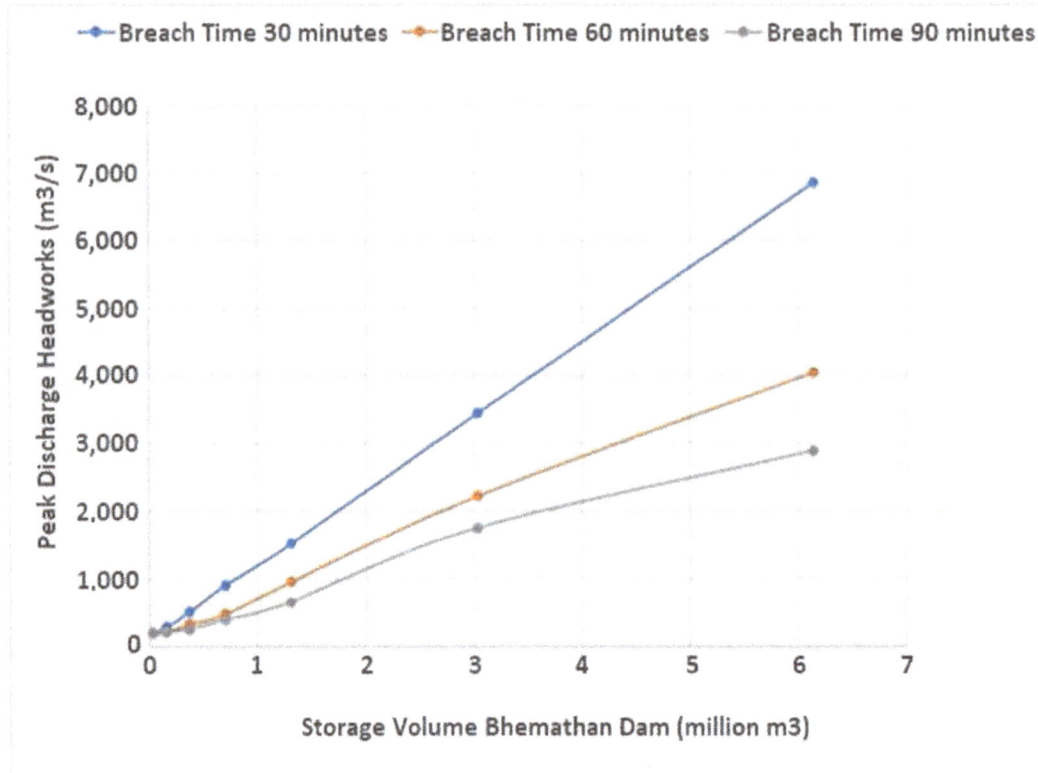

m³ = cubic meter.
Source: Authors.

A range of scenarios arising from the simulation can be considered with event tree analyses. Two scenarios are presented Table 7. The first scenario "Sum (2-7)" assumes simultaneity in all dam breaks as a conservative estimate, while the second scenario "Sum (6-7)" considers the maximum outburst volume from Bhemathan with a future topography, released at the same time as the outburst from a future glacial lake. The total volume under the second scenario could be 2 million m³, and the assumed resulting maximum outburst flow discharge could be 2,200 m³/s.

The breach time or erosion speed is a crucial statistic for converting the storage volume into peak discharges. However, no further investigation on the impacts of this parameter was done in the present study. Consequently, the variety of possible combinations of scenarios remains wide, and their associated uncertainties similarly high. It is noted that every 500,000 m³ of storage would approximately change the outburst flow discharge by 500 m³/s to 600 m³/s if a breach time of 30 minutes is assumed.

Flash Floods from Cloudbursts

Cloudbursts were not observed in the Melamchi catchment in 2021, although large distances between monitoring station means they could have occurred above Melamchi Ghyang or further upstream without being noticed. The catchment is susceptible to cloudbursts due to large temperature fluctuations and local rainfall variability, although its entire 324 km² area is unlikely to be affected simultaneously. Therefore, the potential contribution of cloudburst-like phenomenon in extreme floods needs to be adequately reflected in its hydrological design criteria for multi-hazard risk assessment.

Estimates of the water volume that could be generated by cloudbursts above Melamchi Ghyang or Bhemathan, along with corresponding runoff discharge for various rainfall intensities, were computed in the present study. For this purpose, cloudburst or cloudburst-like scenarios with the following intensities were considered:

- The maximum intensity of 20 mm/hr observed in 2021 using satellite data
- The historical maximum intensity of 65 mm/hr from a cloudburst-like event observed in central Nepal on 19-20 July 1993.[40]
- The threshold intensity of 100 mm/hr used in the definition of a cloudburst[41]

Based on the three scenarios, the total contribution of discharge from cloudburst-like events, assuming uniform rain across the 44 km² catchment area upstream of Melamchi Ghyang, was found to be 244 m³/s, 794 m³/s, and 1,222 m³/s, respectively. Similarly, assuming only the 34 km² catchment area upstream of Bhemathan to be affected, the estimated runoff discharges from the three cloudburst events were 187 m³/s, 606 m³/s, and 933 m³/s, respectively.

Probable Maximum Flood

The standard approach for estimating the PMF is based on statistical analysis of historical monitored data. This approach generally consists of developing a unit hydrograph, feeding it with the PMP disaggregated over hourly or even sub-hourly values, and reporting results as a PMF hydrograph with a peak as a PMF value. As discussed in Section 4.1, application of this approach to the Melamchi catchment using a PMP of 487.2 mm resulted in a PMF estimate of 3,276 m³/s.

---

40   Shrestha, A. 1998. *Estimation of Evapotranspiration by Penman Monteith Equation and Analysis of General and Extreme Rainfall over Kulekhani River Catchment.* Tribhuvan University, Kathmandu, Nepal.

41   Ashrit, R. 2010. *Investigating the Leh 'Cloudburst'.* https://www.ncmrwf.gov.in/Cloudburst_Investigation_Report.pdf

Summary of Maximum Discharge Calculations

The results of the maximum discharge calculations of selected scenarios for the Melamchi catchment are summarized in Table 8. As noted, these calculations are based on several assumptions related to hazards and are therefore associated with high levels of uncertainty.

Nevertheless, based on the results of the scenarios considered, the maximum discharge of the Melamchi River at the MRWDS headworks, i.e., its PMOF, is recommended as 3,500 m³/s. This value is 30% higher than the estimated maximum flood from the 2021 monsoon (2,700 m³/s) and 5.8 times higher than the design flood of 600 m³/s used in the original headworks design. It should be noted that, without a well-defined return period, the recommended figure is inherently highly uncertain and any design of the MRWDS headworks should take this uncertainty into consideration when defining its design criteria.

Table 8: Summary of PMOF Calculations

Hazard	Analysis assumptions	Discharge (m³/s)	Sum of Discharges (m³/s)
Cascades of dam breaches, synchronous	2,789,557 m³	3,099	
LDOF and GLOF, simplified	2,000,000 m³	2,200	3,422
Cloudburst flash floods	100 mm/hour over an area of 44 km²	1,222	
PMF rainfall-runoff	Based on PMP	3,276	
Recommended PMOF		**3,500**	

GLOF = glacial lake outburst flood, LDOF = landslide dam outburst flood, m³ = cubic meter, m³/s = cubic meter per second, PMF = probable maximum flood, PMOF = probable maximum outburst flood.
Source: Authors.

Uncertainties

Various scenarios of possible multi-hazard combinations, such as those in Table 8, can be established from event tree analyses of the Melamchi catchment, but the uncertainty in the PMOF estimate thus derived will be very high. Even with extensive comprehensive monitoring programs in the future, this uncertainty will continue to persist because of the nature of the Melamchi catchment. This is primarily due to its multi-hazards that may interact in unpredictable ways and uncertainties related to the impacts of climate change. Thus, the recommended PMOF should only be used as guidance in establishing the future design criteria for the MRWDS headworks.

Impacts of Climate Change

Projected Impact of Climate Change on Temperature and Precipitation

In the present study, the future climate of the Melamchi River catchment was projected based on global climate models.[42] The model outputs were corrected for biases using robust quantile mapping, and an ensemble of them was used as the projected time series for further analysis.

The models indicated that the average annual maximum temperature for the mid-future (MF), i.e., 2051 to 2075, was expected to rise by 2.6°C compared to the baseline (1992 to 2021), and for the far future (FF), i.e., 2076 to 2100, by 4.3°C. For the dry season (March to May), the MF and FF temperatures were predicted to rise by 2.5°C and 4.2°C, respectively.

The average annual precipitation value projected from the models had an erratic trend, with lower confidence for the FF. They projected the average annual precipitation levels for the MF and FF to increase by 13% and 29%, respectively, and their projected dry season precipitation value also followed a comparable trend. The projected precipitation increase in the post-monsoon season is more than double the increase in the average annual and/or dry season projection.

Projected Impact of Climate Change on River Discharge

Climate models project that the average annual discharge at the Melamchi headworks will increase by 7% in the near future (NF), by 19% in the MF, and by 36% in FF over its baseline value of 12.9 m³/s (Table 2). Such increasing trends in future discharges are consistent with those reported for other watersheds in Nepal.[43]

A hydrological model used to project changes in future discharge shows that the dry season discharge (March to May) of the Melamchi River is likely to increase by 15% in the NF, 22% in the MF, and 29% in the FF over its baseline value of 4.9 m³/s, considering the assumed climate scenario (SSP585[44]). This indicates a reduced risk in the reliability of dry season flows in the climate change context.

For the monsoon discharge (June to September), the daily maximum flood baseline of 89 m³/s is projected to increase to 149 m³/s and 165 m³/s in the MF and FF, respectively. The projected flood discharge changes have a frequency of only 25 years return period, thus reflecting less risk of climate change on floods simulated only from a rainfall-runoff process. However, higher risks are expected from impacts of climate change on glacier dynamics, associated GLOFs, geo-hazards, and associated LDOF events.

---

42   Mishra, V., Bhatia, U., and Tiwari, A. 2020. *Bias-corrected climate projections from Coupled Model Intercomparison Project-6 (CMIP6) for South Asia.* http://arxiv.org/abs/2006.12976

43   Bajracharya, A., Bajracharya, S., and Shrestha, A. 2018. *Climate change impact assessment on the hydrological regime of the Kaligandaki Basin, Nepal.* https://www.sciencedirect.com/science/article/pii/S0048969717337658

44   Jianchu, X. et al. 2007. *The melting Himalayas: Regional challenges and local impacts of climate change on mountain ecosystems and livelihoods.* International Center for Integrated Mountain Development (ICIMOD)

Climate Change Impacts on Glacier Dynamics and Permafrost

Literature establishing a clear link between climate change and cloudburst phenomenon in the Nepal Himalayas is still not available. However, existing literature on this issue indicates that atmospheric heating will increase the frequency and intensity of extreme precipitation events, ultimately raising the likelihood of hazards such as cloudbursts, flash floods, and landslides. Climate change and associated temperature rises may also accelerate the retreat of snow, ice, and glaciers, and contribute to increases in the volumes of glacial lakes that may subsequently burst and generate GLOFs. Depending on their location and topography, GLOFs may have cascading effects and trigger landslides and LDOFs.

Most glaciers studied in Nepal are undergoing rapid mass loss, with reported rates of glacial retreat up to 20 m per year.[45] The existing glacial lakes in the Melamchi catchment are very small, and associated risk in their current form are considered low. As such, GLOF risks were not modelled in the present study. However, the risks from new glacial lakes emerging due to melting glaciers is included in the estimation of the maximum discharge calculation of the river (Section 5.6).

A warmer climate may also reduce the extent of permafrost that covers an area of approximately 9.2 km² in the headwater of the Melamchi catchment, considering a permafrost line of 4,200 masl. The subsequent thawing of water stored as ground ice in the permafrost region can exacerbate geo- and water-related hazards and risks, such as slope instability, erosion processes, and floods/debris flows, in the catchment. It can also change hydrological processes such as water storage in surface reservoirs including lakes and wetlands. It can affect hydrological connectivity and surface water-groundwater[46] interaction, resulting in an increase in groundwater storage and recharge.[47]

---

45   Yao, T. et al. 2020. *Third pole climate warning and cryosphere system changes.* https://public.wmo.int/en/resources/bulletin/third-pole-climate-warming-and-cryosphere-system-changes

46   Connon, R. et al. 2014. Changing hydrologic connectivity due to permafrost thaw in the lower Liard River valley, NWT, Canada. *Hydrological Processes*, 28 (14): 4163-4178.

47   Bense, V.F., Kooi, H., Ferguson, G. and Read, T. 2012. Permafrost degradation as a control on hydrogeological regime shifts in a warming climate. *Journal of Geophysical Research: Earth Surface*, 117 (F3).

Figure 42: Extent of Permafrost in the Melamchi Catchment

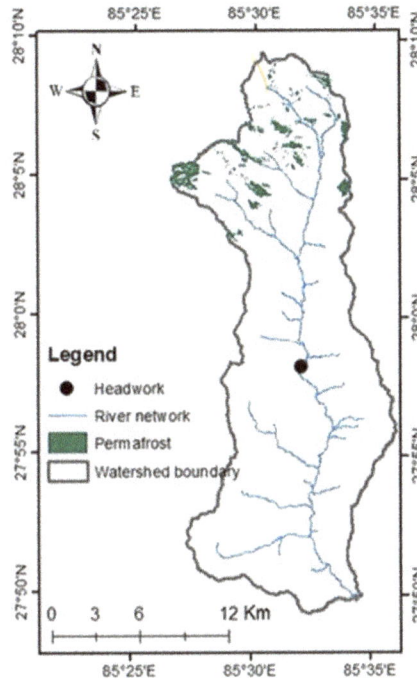

Note: Does not include bedrock areas.
Source: Authors.

Importantly, the active layer that holds the ice together at critical joints can also disappear as temperatures rise. As the ice cohesion diminishes, erosion can penetrate deeper, and talus slopes can produce larger debris flows. Permafrost thawing associated with climate change will thus increase the risk of LDOF as well as floods and debris flows as it releases more sediment-laden mud-like flows and exposes more areas to geo-hazards.

Assessment Uncertainties

Climate change impact assessments contain uncertainties in terms of the representation of local models. Flow hydrographs depend upon various factors such as the geomorphological movement of materials including rocks and soil, and the potential heights of landslide dams. The mode of dam failures and time required for the water to be released and travel to a point of interest, such as the headworks, also need to be considered.

Quantitative estimates of the implications of climate change in glacier dynamics and permafrost retreat for the Melamchi River catchment could not be provided due to a lack of adequate observed data. As such, given the uncertainties in defining future maximum flood discharges, the recommended PMOF value of 3,500 m^3/s is retained. Further, larger regional models are recommended for better climate change impact assessment.

6. STABILIZATION AND MITIGATION MEASURES

The extensive damage to the MRWDS headworks caused by the 2021 debris flood events has severely impacted the water supply in the Kathmandu Valley. To avoid future impacts to the valley's water supply, three categories of mitigation measures were considered to reduce the hazard intensity of extreme debris and flash floods in the Melamchi catchment.

- **Monitoring** through EWS for timely response
- **Management** of hazards when they occur using information from EWS
- **Stabilization** of the catchment to prevent or minimize hazard occurrence

Given the condition of the Melamchi River catchment, the natural geomorphological processes that led to the 2021 events, including landslides, sediment deposition and erosion from flood hazards, etc., will continue in the river catchment over the foreseeable future. However, as the number and scale of potential lateral sediment sources in the catchment are too high for any meaningful sediment management strategy, catchment stabilization measures were not examined in the present study. Therefore, the study focused on effective monitoring and management of hazards to reduce their impacts on the headworks.

Although intended to minimize damage to the Melamchi headworks, the proposed mitigation measures, including micro-scale stabilization measures against landslides, will directly benefit the local population by protecting settlements and securing access roads and critical infrastructure. For the long-term, an integrated river basin management (IRBM) plan is recommended for sustainable development in the catchment.

Slope and Landslide Monitoring

Landslide Early Warning Systems

Installation of landslide early warning systems (LEWS) in the river catchment is an effective way of increasing preparedness and triggering timely responses. These systems can provide advance warnings about landslide events that could cause nat ural dams, outburst flooding or runouts near headworks structures. This allows suitable and timely response actions, such as more detailed monitoring and contingency measures, to be initiated.

LEWS are proposed at the following sites that have active landslides or have been identified as slopes at risk of failure:

- Bhemathan
- Melamchi Ghyang and other critical sites between Bhemathan and the headworks
- Headworks left bank landslide

Recommended Types of LEWS

LEWS can consist of any one or a combination of the following monitoring tools:

- Remote sensing with satellite image analysis
- Inclinometer
- Piezometer
- Survey of landmark points on slopes using stable control points
- Monitoring of tension crack[48]
- Light Detection and Ranging (LIDAR) scanning

Remote sensing using Interferometric Synthetic Aperture Radar (InSAR) radar satellite for land displacement analyses of slow-moving landslides is recommended for characterization of landslide susceptibility. Other Earth Observation products for remote sensing should also be explored for monitoring of slope instability through hot spots detection, monitoring of topography change detection (debris flow accumulation, gully erosion, etc.); and satellite monitoring of the extent of glaciers, glacier lakes, and snow cover.

Inclinometers are recommended to determine the magnitude, rate, direction, depth, and type of landslide movement, and thus diagnose the cause, expected behavior, and eventually the remediation of a landslide. Of the various types of inclinometer probes, a traversing probe is recommended to record the inclination at predetermined intervals and establish a continuous profile of the casing shape. Newer micro-electromechanical system digital inclinometers are also recommended to provide real-time online data. In addition, the Shape Accel Array system, which provides real-time data for the full borehole length, may be considered.

Figure 43: Using a Long-Range Light Detection and Ranging (LIDAR) Scanner and Inclinometer to Detect Motion

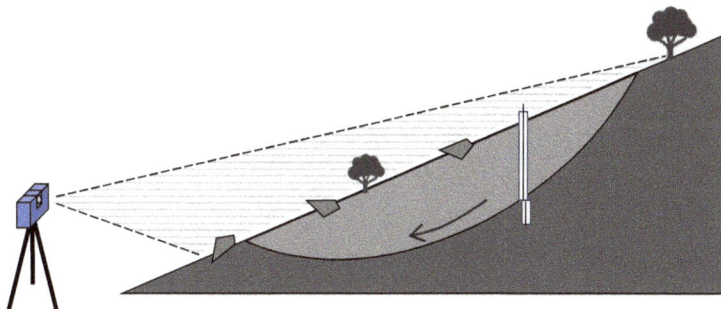

Note: Image shows long-range 3D LIDAR scanner on the left, inclinometer on the right.

Source: Authors.

---

48   Cracks or open fractures that develop in a slope when the inclination angle of the slip surface is steep. They can also be found along the flanks outside the main landslide zone. Scarps, which are generally found near the top of a landslide in the zone of extension, generally begin as tension cracks.

Piezometers should be installed near sliding planes to measure and monitor changes in pore water pressures. Typically, vibrating wire piezometers are preferred.

Survey of landmark points is recommended for accurate surface measurement of slopes near the headworks. The surveys may employ total stations for manual measurements of fixed targets on slopes or a stationary three-dimensional LIDAR scanner or a simple mounted camera that takes photos at given intervals from the same spot can detect movements of larger objects such as trees and rocks on the slope.

Monitoring of tension cracks at or near the crown of landslides, such as those at left bank of the Melamchi headworks, or elsewhere, should focus on the pattern, length, and width of the cracks along with the vertical offset between their two sides. Movements along cracks may be simply measured by firmly fixing steel or timber pegs on each side at selected locations and periodically measuring changes in their separation. The frequency of monitoring can be weekly if there is no significant change, and daily if noticeable movement is detected. The measurements should be documented to draw inferences on the magnitude and direction of slope movement.

Figure 8.1: Tension Cracks above the Melamchi Left Bank Landslide

Source: Authors.

LIDAR surveys of landslide areas should be conducted at regular intervals, or as required, to produce high resolution maps that can be used to identify displacement locations, magnitudes, and velocities. Three options for LIDAR surveys can be considered: aircraft-borne for large areas, drone-borne for medium areas of 1-2 km, and tripod-borne for near-field surveys for areas up to 100-200 m.

Plans for LEWS at various sites and stretches of the Melamchi catchment were proposed in the present study considering the nature of the landslides at these locations (Table 9). At Bhemathan, satellite remote sensing and inclinometers were recommended to monitor the stability of the deposit's end slope that is undergoing progressive backward erosion. Because of their limited accessibility, the 11 landslides analyzed for LDOF between Bhemathan and the Melamchi headworks were recommended to be monitored only with satellite image analysis. At the headworks landslide, a combination of LEWS was recommended to monitor further movements, especially during monsoons.

Table 9: Landslide Monitoring in the Melamchi Catchment

Monitoring System	Location	Quantity
Bhemathan Deposit		
Remote sensing with satellite image analysis	Entire deposit	1
Inclinometer	2 at center, 1 at left, and 1 at right	4
Sites between Bhemathan and Headworks		
Remote sensing with satellite image analysis	Bhemathan to Melamchi headworks	1
Headworks Left Bank Landslide		
Remote sensing with satellite image analysis	At headworks	1
Inclinometer	2 above and 2 below tension cracks	4
Piezometers for pore pressure measurements	Near sliding plane	10
Survey of landmark points on the slope from stable control point	Lower part of slope	10
Tension crack monitoring: change in number, length, and width of each crack	Between settlement and landslide	30
LIDAR scanners, either stationary or using drones or aircraft	Entire slope	1

Source: Authors.

The proposed LEWS plan for the headworks landslide will be further developed in phase 2 of the study.

Management of Landslide Hazards

EWS are only relevant when adequate responses, protocols, and procedures are in place and can be activated if alerts are triggered. At each site in the Melamchi catchment where monitoring is recommended, EWS technology should be used in conjunction with local DRM responses customized to cater to needs of the site, the nature of its slopes, and its vulnerabilities.

At each site, those responsible for receiving early warnings must have immediate access to a comprehensive DRM plan, both online and in a hardcopy format in case the internet and power fail. At a minimum, this plan should contain: a) hazard trigger definition (when to act); b) alarms to be activated, c) local DRM authorities to be alerted, and d) people in their own organization to be alerted. The DRM should as a minimum establish the following short-term actions to be performed in response to a disaster:

- Identify local responders and establish a chain of command to approve alert messages and to communicate in an emergency or during a disaster.
- Encourage all people in the risk-prone area to sign up for emergency alerts.
- Practice alert messages and communications.
- Identify all communities and households at risk and make evacuation plans.
- Evacuate immediately, if told to evacuate.
- If an earthquake or prolonged rainfall increases landslides risk, intensify further monitoring of hazardous locations such as Bhemathan, Melamchi Gyang, and the headworks.
- Use land-based (vehicles) or aerial technology (drone) to verify the risks if events are hazardous.

In general, the removal of newly formed landslide dams or the drainage of newly formed lakes or reservoirs due to blockage of the Melamchi River or its tributaries will neither be practical nor recommended due to difficulties in accessing remote sites. Instead, more intense monitoring is recommended to help quickly understand the cause and assess the hazard escalation risk.

Site-Specific Management Plans

Landslide hazard management plans were prepared for each site in the Melamchi catchment where slopes are at risk of failure and have been recommended for monitoring (Table 10). At the Bhemathan plain, a temporary blockage of the river due to debris could be removed in a controlled manner such as by blasting, provided that extended monitoring of the blockage confirms its feasibility.

Of the five landslides between Bhemathan and the headworks that are considered critical, the ones furthest up and downstream could be monitored with additional EWS that could be activated following a primary alert. At the headworks landslide, both the large river discharge seen during floods, and small or no discharge due to upstream river blockage, could also be monitored.

Table 10: Extended Monitoring at Different Sites in Case of Early Hazard Warning

Extended Monitoring after Early Warning	Locations	Quantity
Bhemathan Deposit		
Water triggered sensor (level gauge)	Selected sections on plain downstream of end-slope toe	Minimum of 2 sensors
Subsequent close-up aerial orthophotos of events registered through EWS, by drone or helicopter, weather permitting	Entire plain	1 set of photos, convertible to DEM
Sites between Bhemathan and Melamchi Headworks		
Water-triggered sensor	Upstream of most upper critical landslide	2
Dry-triggered sensor	Downstream of most downstream critical landslide	2
Headworks Left Bank Landslide		
Water-triggered sensor	River section at headworks	1
Dry-triggered sensor	River section at headworks	1

DEM = digital elevation model, EWS = early warning system

Source: Authors.

Disaster Management Plan

A DRM or disaster response plan should form an essential part of the MWRDS's control system. The DRM should contain all emergency operation procedures (EOP) and response plan, and be developed in line with prevailing guidelines and standards in Nepal. In particular, the DRM should include

- Links to EWS;
- Mapping of potential hazard risks and trigger thresholds for alerts;
- Impacts, responses, and operation continuity analyses;
- Communication plans; and
- Protocols for testing and updating the disaster management plan.

Real-Time Water Level Monitoring Sensor

Water level sensors which measure the distance from sensors mounted above water and down to the water surface are recommended, as in-stream sensors may easily be destroyed in high flows or debris floods. Ultrasonic or radar-based level sensors should be used, and they should be equipped with a solar panel and battery, local data storage, and appropriate data transmission equipment.

Secondly, control cross-sections must be established where i) the water flow is confined to a certain width so that the measurement beam from the sensor will always be above water, and ii) a rack can be established across the river for mounting the sensor. Ideally, this rack can be combined with a walkway to benefit local residents.

Close-Up Orthophoto

If EWS reveal significant changes to the slopes of the Bhemathan plain, a rapid overview of the area using a camera mounted on either a helicopter or drone is recommended. A camera carrier that can be used during rainfall is required and should be available for mobilization with a few minutes' notice. Ground control points (GCPs) must be established beforehand near specific sites where orthophotos are needed.

Stabilization of Landslides

Mitigation solutions that help eliminate or reduce the risk of landslide occurrence can also offer broader cost benefits such as protection of local property and infrastructure, including roads, bridges, power lines, and water supplies. The following mitigation measures are considered for each vulnerable area:

- **Water drainage** of slopes, using surface water contour drains, catch drains, drainage canals along with perforated pipes, and drain holes for groundwater
- **Slope grading**, such as terracing, benching, flattening, debris catch, check dams, bamboo fencing, selective and controlled material removal, etc.
- **Toe protection** using retaining and gabion walls, revetments and other riverbank erosion protection, enlargement of slope toe, etc.
- **Fixing slopes** with forestation, vegetation, rock bolts, shotcrete, geogrid, geofabric, geotextile, mulching/ground cover, and soil nailing
- **Check dams** for controlling sediment and debris from smaller annual floods.

Stabilization of Bhemathan End Slope

The Bhemathan end slope (Figure 45) is not yet stable, and there is a possibility of its further erosion in future monsoons. In order to control this erosion, a low-height (about 3 m high), gabion-type check dam was considered across the entire 35 m river width. Located about 700 m downstream of the end-slope, this dam would form a flatter end slope with the help of sediments arrested from smaller debris flood events and would thus provide stability to the Bhemathan deposit.

Figure 45: The Bhemathan Knickpoint Seen from Downstream

Source: Authors.

Figure 46: Melamchi River Profile from Bhemathan Knickpoint Based on DEM

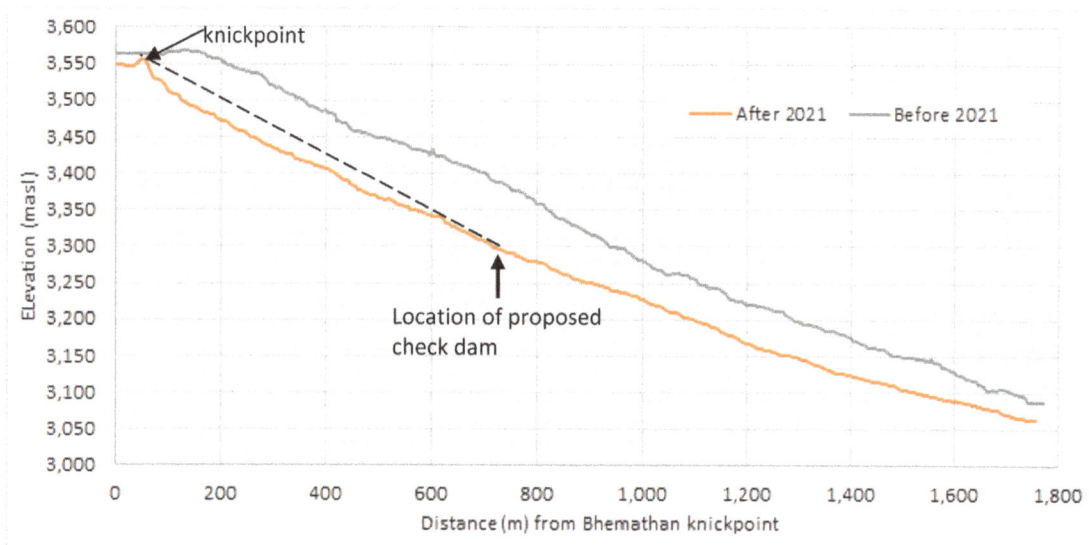

DEM = digital elevation model, m = meter, masl = meters above sea level

Source: Authors.

Further analysis showed that the check dam would add value for small annual floods, with additional benefits besides protection to the headworks. However, it would be swept away in catastrophic floods with forces comparable to those in 2021. In addition, construction of this dam would be difficult because of the remoteness of the site and its difficult access. As such, this idea was dropped.

Stabilization of Landslides between Bhemathan and Headworks

Melamchi Ghyang is the furthest upstream village in the catchment, almost halfway between Bhemathan and the headworks. This river stretch also has other villages, some of which are located above existing landslides.

While it may not be cost-effective from a project perspective to protect the Melamchi River and the headworks from all landslides, it is important to minimize the risk to villages from selected landslides. Local and nature-based slope protection, coupled with proper drainage measures, will provide some benefit, and measures to secure roads and bridges will help provide access to future monitoring stations and increase management options for the upper reaches.

Water drainage is essential for slopes and undertaking local feasibility studies for low-cost solutions for a few slopes is recommended. Slope grading may be difficult in some areas, but it is crucial that settlement is avoided near the top of potentially unstable slopes to reduce loads on the slope and thereby minimize risks of its failure. Regulation of buildings, roads, and other land uses is also important to minimize human-induced disturbances to such slopes.

Gabion-retaining structures, which are cost-efficient and can readily be constructed locally, are recommended to mitigate localized erosion and small slope instabilities. However, these will not work for deep-seated landslides. Low-height gabion check dams across smaller tributaries can be used to effectively control their sediment and debris contribution to the main river during minor floods.

Bioengineering solutions are recommended to fix slopes in areas where human life, property, and infrastructure are at risk. Suitable where bedrock is not exposed, these low-cost measures work well under normal conditions but may not provide sufficient safety against hazards under extreme events. Vetiver is one of the plants used for slope stability due to its very long roots. Slope-fixing solutions such as water drainage combined with ground cover, toe protection with shotcretes, rock bolts, and/or soil nailing, can also be considered.

Figure 47: Typical Cross-Section of Gabion Wall to Be Constructed at Landslide Toe

Source: Authors.

Stabilization of Headworks Left Bank Landslide

The landslide at the Melamchi headworks site is critical to the functioning of the MRWDS, and further studies on it will be undertaken in Phase 2 of the study. A separate report prepared by the TA team[49] recommends the following short-term stabilization interventions for this slope to prevent further instabilities and monitor its movements:

1. Establishment of an efficient catch drain with sufficient capacity along the existing road just above the landslide.
2. Interception of any streams and gullies above the road and their diversion into a road drain, including management of the right-side gully.
3. Safe release of the road drain south of the headworks left bank landslide.
4. Sealing of tension cracks at the top of the landslide to prevent rainwater from entering them and saturating the slope soil.
5. Establishment of benchmarks at key hot spots on the landslides and their monitoring on a weekly, or as needed, basis during monsoon to track surface gravitational movements.

---

49   ADB and MWSDB. 2022. *Damage assessment of Melamchi Headworks and Hazard Mapping of the Catchment, TA6596-NEP.* Note on Urgent First Priority Work, 31 March 2022.

6. Careful removal of large, loose, and fractured boulders in the middle and lower parts of the landslide that are prone to rolling down during the monsoon.

In addition, the following long-term stabilization interventions were recommended based on best practices and in line with current Nepalese regulations:

1. Maintenance of the established drain systems and crack seals, and expansion of the drainage system with more canals (ditches lined with impermeable material such as geomembrane or plastic and filled with rocks).
2. Gradual removal of loose soil and boulders starting from the top of the slope using excavators until sound bedrock or a flatter stable slope is achieved.
3. Exploratory borehole drilling, with full core recovery using double or triple core barrels, at least 10 m into the bedrock, followed by field and laboratory tests to determine the engineering properties of the slope material.
4. Installation of inclinometers to monitor potential slope movements at depth.
5. Installation of piezometers to measure pore water pressures near the slip surface.
6. Updating of existing slope stability analyses using monitoring data from inclinometers and piezometers, and development of slope stabilization measures considering stable angles with intermediate berms, soil and rock reinforcement, and surface and subsurface drainage.

It is noted, however, that complete stabilization of the slope in a cost-effective manner may still be elusive because of its unfavorable geology (dip slope, Section 3.1), deep-seated failure surface, steep inclination, and fragility. Hence, the slope will persist as a constant concern for the headworks site.

IRBM Planning

This is used to analyze, plan, adjust, and develop the use of land, water, and other resources to enhance long-term hot spoteconomic and environmental sustainability. It involves myriad disciplines including agricultural development, flood and drought management, urbanization, and poverty reduction, and entails cross-disciplinary coordination of activities across all sectors.

Typical IRBM phases over a 5-year period include i) definition and demarcation of planning areas; ii) initial baseline description and analyses of each sector and prevailing natural resources; iii) hot spots projects emerging from such baseline analyses; and iv) IRBM planning including stakeholder sensitization and participatory plan development, alongside long-term implementation.

As landslides and slope protection impact many different sectors, an IRBM approach to river catchment development is recommended. Hot spots projects such as slope stabilization in connection with road network development are priority areas and should be achieved early in the process without compromising the overall development objectives.

River Channel Monitoring

Comprehensive monitoring of both the river and the wider catchment is central to effectively mitigating the myriad risks posed by events including potentially catastrophic floods and landslides in the area. Hydrology and river morphology monitoring can capture real-time data needed to support an EWS and enable swift and effective response to disasters. In addition, it generates information needed to prepare effective disaster response plans and implement disaster relief strategies to cope with the aftermath of flood events.

This section outlines the requirements of various monitoring methods and considers river morphology stabilization options.

Hydro-Meteorological Monitoring

There are three meteorological stations within the 161 km² area of the Melamchi catchment upstream of the headworks. The meteorological station density in this area is one station per 54 km², which is above the WMO's ideal density of one station for each 100-250 km². However, all three stations are located on the eastern side of the catchment and no stations are present in the upper catchment area where most hazardous debris events are initiated. Furthermore, the stations record daily instead of hourly data and do not measure temperature or other meteorological variables.

Only one hydrological station covering a little more than half of the total catchment was present at the time of this study. No stations are present in the downstream part of the catchment and importantly, there are no hydrological stations at the MRWDS headworks site.

As the MWSP is a lifeline solution for Kathmandu residents, high priority should be given to hydrological and meteorological monitoring in the catchment. To design effective EWS, automatic stations that can monitor hydrometeorology should be installed in the catchment to capture cloudburst-like phenomena and associated short-duration river discharge on an hourly basis. The issue of both spatial and temporal measurements and the need for monitoring more meteorological variables should be addressed by Nepal's DHM.

River Morphology (Topography) Monitoring

Monitoring sediment processes in hazardous areas requires a multi-staged interdisciplinary approach that generates data sets at different time scales. In the Melamchi River, the Bhemathan knickpoint slope and the landslides near the headworks and further upstream are the most hazardous areas to be monitored.

Earth Observation technologies and ground-based instruments tracking displacement should be linked to rainfall monitoring. This would allow a threshold rainfall value to be indirectly used to more intensively monitor the initiation of fast-moving debris. Both rainfall and satellite image data could then provide integrated information of the Melamchi River's morphology.

Regular monitoring of the river channel's geometry–planform, cross-section, and longitudinal profile and associated processes, including erosion and deposition, is proposed. This would trace the dynamical fluvial geomorphological behavior of the river, including factors such as riverbed aggradation and degradation, head cutting, gravel bar migration, bank erosion, floodplain sedimentation, and debris accumulation at bottlenecks. It would also facilitate the sustainable operation and maintenance (O&M) of structures in the river, such as the headworks, check dams, bridge abutments, and toe protection works.

Table 11: Requirements of Morphological Monitoring of Melamchi River

Monitoring Location	Monitoring Objective	Instruments/ Methods	Monitoring Matrix	Monitoring Frequency
• Bhemathan deposits, knickpoint, and reaches 2.2 kilometers downstream of knickpoint • Nakote • Melamchi headworks to Timbu Bazaar	• Detect the dynamics of channel planform, change, and erosion/ deposition on Bhemathan deposits • Detect the trend of knickpoint erosion	• Field observation • Camera • Google Images • Aerial photos • Drone survey • Cross-section/L-profile • LIDAR scanners	• Planform, corridor/active/ wetted width • Spatial distribution of erosion and deposition • Erosion and deposition volume	• Annually (Divided into phases, first phase could be for next 5 years)
• Melamchi headworks to Timbu Bazaar	• Detect the dynamics of channel planform, change and erosion/ deposition, and bank cutting	• Field observation • Camera • Google Images • Drone survey • Cross-section/L-profile	• Planform, corridor/ active/ wetted width • Spatial distribution of erosion and deposition • Erosion and Deposition volume	• Annually (Divided into phases, first phase for next 5 years)

LIDAR = Light Detection and Ranging.

Source: Authors.

River Monitoring

Due to the prevailing conditions in the Melamchi catchment and their importance for MRDWS, installation of three automatic stations with 1 to 5 minute temporal resolution is recommended for monitoring hydrology as well as meteorological/climatic variables. A real-time water level monitoring system should be comprised of a water level sensor, signal processor, transmitter, receiver, alarm siren, and batteries with solar panels. When the water level, rainfall, or sediment reaches a pre-defined threshold risk level, a siren should automatically produce a warning signal.

For long-term, large-scale monitoring of the river's plan form and overall profile, the establishment and use of a remote sensing system with satellite image analysis is recommended. However, closed-circuit television (CCTV) cameras are proposed for the short-term monitoring of river morphological activities. For instance, a high-resolution digital CCTV camera could be fixed on top of the hill near the Bhemathan deposit and focused on the knickpoint area to monitor changes in the bed topography and plan form and to obtain information on water surface and flow depth.

The use of drones is proposed for annual river bathymetry surveys during the dry season when much of the riverbed is exposed. The remaining wetted cross-section outside of pond areas will typically be less than 1 m deep and does not need to be accurately surveyed. As technology develops, LIDAR may become more price competitive.

The proposed bathymetric surveys can be performed with fixed-wing drones with red, green, and blue cameras, with a minimum of three flight lines along the river's course and eastern and western valley slopes between the headworks and Bhemathan. The flight time should typically be 1 hour per 5 km river stretch. In addition, GCPs must be established along stable banks of the river channel, with one GCP for every 500 m and a higher density near the headworks. The topography data from the survey should be processed using GIS, and changes in the river's longitudinal profiles, cross-sections, annual erosion, and deposition patterns shall be extracted by comparing subsequent topography maps.

Table 12: Hydro-Meteorological, Morphology Monitoring Parameters

Location	Description
Hydrological stations: River water level, discharge, and suspended sediment monitoring	
Nakote, at or nearby existing station (Priority 1)	• Parameters: water level, discharge (based on rating curve, to be developed by new-Bed-Level survey), and total suspended solids sediment • Temporal resolution: 1-5 minutes • Data transmission and acquisition: local storage in SSD card plus real-time transmission to server. Data transmission technology should consist of radio link for regular communication, mobile network (3G or 4G), with satellite system for backup and emergency use • The station equipped with all necessary instruments, power backup, communication channel, etc, with adequate redundancy in all components
At suspension bridge located at the gorge between Talamarang to Bhattar Phaat, about 30 km downstream of MRWDS headworks (Priority 1)	
At MRWDS headworks (This will be a part of the headworks and be considered in Phase 2) (Priority 2)	

continued on next page

Table 12 *continued*

Location	Description
Meteorological station: Meteorological and climatic data, precipitation, temperature etc.	
Bhemathan (Priority 1)	• Parameters: Rainfall, temperature, humidity, wind speed and direction, solar radiation, and soil moisture • Temporal resolution: 1–5 minutes • Real-time oblique photography, integrated with automatic rainfall station • Data transmission and acquisition: local storage in SSD card plus real-time transmission to server. Data transmission technology should consist of radio link for regular communication, mobile network (3G or 4G) and satellite system for backup and emergency use • The station shall be equipped with all necessary instruments, power backup, communication channel, etc, with adequate redundancy in all the components
Headworks, Nakote area, or suitable site to be identified (Priority 1)	
Melamchi Bazaar (Priority 2)	
River morphology monitoring system: Riverbed and floodplain topography	
Installation of CCTV cameras	• Remotely operated and lot-powered CCTV camera • Erection of shed and arrangement of security, power, connectivity etc.
Establishment of GCPs from the headworks to Bhemathan	• GCP installation for every 500 m: identification of permanent fix point and marking (bedrock outcrop) • One-time survey of each GCP using total station to obtain position and elevation of each
Drone survey once a year during dry season (December) For landslide movement detection with LIDAR, several times over the year	• Fixed-wing drone, with DGPS positioning system, and RGB camera • Photogrammetric model preparation using appropriate software • Alternatively with LIDAR scanners installed • Data analyses using GIS for extracting profiles, erosion and deposition levels and volumes
Link to remote sensing system	• Earth Observation for detection of river plan form changes

CCTV = closed-circuit television, DGPS = differential global positioning system, GCP = ground control points, GIS = geographic information system, LIDAR = Light Detection and Ranging, MRWDS = Melamchi River Water Diversion Subproject.

Source: Authors.

Management of Flow, Flood, Sediment, and Debris

Short- and long-term measures are suggested to better manage drought, floods, and debris specifically related to river hydraulics and morphology (Table 13). This provision should start with the definition of a list of adverse scenarios that may occur during floods or droughts, and which should be continuously updated in a disaster management plan as well as an O&M manual with EOPs.

Table 13: Management Options against Adverse River Morphology Events

Adverse Event	Impact	Management Response
Drought over longer periods (days, weeks)	Insufficient water available for diversion to the MRWDS over a longer period	- Diversion of flow before the drought to a buffer storage - Temporary shift to an alternative raw water source when buffer storage is exhausted
Low flow over a short time (hours) due to river blockage by debris	Insufficient water available for diversion to the MRWDS	- Use of buffer storage - River inspection to determine reason and location of flow blockage, and initiation of clearance
Unforeseen contamination of raw water, from seepage to river upstream, deliberate malicious actions etc.	Unsafe raw water, need for further treatment at water treatment plant	- Intensified water quality monitoring upstream of the headworks - River inspection for reason and location for contamination
Flooding over longer time during monsoon (days, weeks)	Flooding of equipment, vehicles, instruments, assets used for O&M, predictable water levels	Remove equipment, vehicles etc. temporarily from flood prone areas as per O&M EOPs
Flash flooding over short period (minutes, hours)	Flooding of headworks area, unpredictable water levels	Raise alert of potential disaster and intensify debris monitoring
Sediment and mud floods as part of flooding	Blocking of gates and other assets by sediment bars, and/or scour holes jeopardizing structural integrity	Removal of sand and gravel bars after event, fill in scour holes in riverbed with earth-moving equipment on standby such as excavators
Debris flooding as part of flash floods	Damage to headworks structures and instruments	- Temporary closing of water intake until debris flood has passed. - Removal of debris after the event

EOP = emergency operation procedure, MRWDS = Melamchi River Water Diversion Subproject, O&M = operation and maintenance.

Source: Authors.

Corresponding to the design discharge of 2 m³/s for the MRWDS, a 1-hour shortage in water supply would require a buffer storage of 7,500 m³, while a one-day outage would need 173,000 m³. This buffer can be provided in the 26 km water tunnel and its Gyalthum and Sindhu adits.

River inspections could be performed at low cost by assigning residents with the responsibility for a 500 m reach of the river upstream of the headworks. Inspections of hard-to-reach areas could be carried out by drone, depending on weather conditions.

If severe debris floods are followed by deposition at key locations such as in front of the intake, adequate equipment should be available to quickly remove the debris and thus shorten the recovery and outage period. Standby equipment with fuel and spare parts is recommended on either bank of the headworks and at other critical locations.

Stabilization of River Channel

Permanent structural mitigation measures such as river training works can reduce or even eliminate the impact of hazard events. These constructions, which are designed to stabilize, confine, and guide the river flows, are considered at three locations:

- Around the Bhemathan plain
- Between Bhemathan and the headworks
- Around the headworks

Bhemathan Plain

During monsoon seasons, the channel course over the Bhemathan plain could shift from the left to the right corner of the knickpoint (looking downstream). This could result in erosion of significant amounts of loose material present on the right side of the knickpoint. The knickpoint itself acts as an erodible sill with steep downstream slope, over which very high flow velocity and shear stress can lead to channel scour and bank erosion. However, as these erosion processes at the plain and knickpoint will continue for many years, construction of any dike structures to change the natural development pattern is not recommended.

The point of overtopping at the knickpoint appears to be a bottleneck which could be blocked by floating debris. This could gradually turn into a new dam which could be breached to cause another flash flood. An option is to blast the fractured rock which forms the bottleneck to ensure more controlled erosion and future degradation of the knickpoint. A counterargument against blasting is that organic debris could quickly overtop, or the sill could widen because the flanks consist of erodible material. Here too, no structural interventions are recommended owing to uncertainty in their effectiveness.

Between Bhemathan and Headworks

Catchment-wide river training works to manage possible floods are unfeasible due to their cost, uncertainty about the future river channel configuration, and concerns over the impact of morphological conditions including floods and debris at the headworks. However, if an IRBM plan which considers multiple interests and stakeholders alongside the MRWDS is adopted, some measures could be tied in with the residents' calls for protection of vulnerable riverbanks, and reconstruction of roads and walkways across the river. These could become priority hot spots projects, and river training locations could be decided through field observations and local consultations. Locations that should be considered for river training work are the outer bend or left bank of the river at Nakote, and the reach between the Sarkathali suspension bridge and the narrow gorge upstream of the headworks.

Around the Headworks

A number of options for stabilizing the river channel in the headworks area can be considered to protect the headworks structures from further damage. Retaining structures may be built along the toe of the left bank landslide and along the right riverbank to channelize the river and thus reduce risks of landslide aggravation, right bank erosion, and even deposition in the area during floods. These options will be further elaborated, evaluated, and finalized in Phase 2 of the study. However, the technical and cost-effectiveness of these measures may be difficult to guarantee under the hazard conditions that exist in the catchment and around the headworks site.

Preliminary Headworks Design Considerations

Final decisions about river training works are closely related to the headworks design and will be fully assessed in Phase 2. The following preliminary observations and considerations of the existing headworks are based purely on the interpretation of field observations and diagnostics of causes and impacts.

Landslide

Three possible scenarios are possible for the headworks left bank landslide which has been active since 2020. It may stop moving and stabilize by itself, continue to slowly progress or creep over many years, or undergo sudden slope failure. Even if it stabilizes, the landslide will continue to pose a major threat to the headworks and its downstream.

In-Stream versus Off-Stream Structures

The previous headworks structures built within the river channel were directly exposed to the enormous impact of the debris floods, which would have severely damaged almost any design. For better protection against such elements, future headworks structures should be placed away from the main channel, and approaches such as side intakes, riverbed infiltration, and flow diversion should be considered.

Sediment Deposition Zone

The present headworks area is just downstream of a steep and narrow gorge, at a location where the river channel and floodplain widen out. The river also takes a 90-degree turn just downstream of the headworks. The river planform–or horizontal geometry–is therefore prone to sedimentation and riverbed aggradation. The tendency toward sediment deposition at the headworks could be averted through extensive river training work, river straightening, and compartmentalization of the river's cross-section profile.

Headworks Location

New sites for the headworks will be considered and reviewed in Phase 2 by the TA team and the MWSDB, based on technical and economic assessments. The repurposing and reuse of the existing headworks structures will also be considered based on the results of this study and additional risks caused by the left bank landslide. Phase 2 will assess the same factors that influenced the selection of the original headworks area, such as valley slope, lateral water inflow, geology, infrastructure access, as well as environmental and social constraints.

Headworks Redesign and Intermediate Solutions

While the MRWDS headworks were destroyed in 2021, other parts of the water supply project such as the water diversion tunnel are working as planned. Conceptual design solutions in Phase 2 should consider the need for robustness and resilience of future infrastructure. They should also explore means of integrating intermediate water intake solutions with the existing infrastructure till the headworks are fully recovered and rebuilt.

Summary and Recommendations

Summary

This report examined the factors that triggered the 2021 hazards, assessed their impacts on the MRWDS headworks and the Melamchi catchment, and forecast future risks. It also proposed new criteria for estimating the maximum flood in the catchment-the PMOF-that integrates the influence of the catchment geology geomorphology as well as climate change with the rainfall-runoff hydrological procedures adopted conventionally. It further explored options for monitoring and managing the river catchment and headworks area, and suggested short-, medium-, and long-term measures for stabilizing the river and thereby reducing risks to the catchment and the headworks.

These findings and recommendations will be further examined and used in Phase 2 that will primarily focus on repurposing and rehabilitating the MRWDS headworks to ensure year-round bulk water supply to the Kathmandu Valley. This study will explore various strategies, including design philosophies and approaches, to enhance the resilience of the headworks structures and reduce its vulnerability to future hazards. Considering the challenges involved in treating the catchment, the study will investigate means of introducing redundancy in the water diversion system to ensure uninterrupted water supplies. Phase 2 studies started in September 2022 and were completed in a year.

Recommendations

This comprehensive study recommends several interventions at the project, catchment, and regional levels for addressing the causes and effects of the hazards that were witnessed in the 2021 events. Its main takeaways, which are equally applicable to similar catchments in the region, are summarized at three levels:

Project-Scale Interventions

Rehabilitation of the MRWDS headworks infrastructure is feasible but work needs to be staggered to avoid jeopardizing the flow of water to the capital.

Urgent short-term measures for restoring the water supply should include clearance of debris at the headworks site, hydrological surveys, and improved sediment management at the temporary water intake. The headworks landslide should be monitored, and water drainage work undertaken to help mitigate the risk it poses. These measures should be carried out immediately.

Medium-term solutions should include measures to protect the water intake and ensure water supplies until long-term solutions are put in place. These measures include installation of sediment screens in front of the water intake to remove large sediments from the diverted water, rehabilitation of the desander and other structures to prevent finer sediments entering the water tunnel, and the stabilization of the toe of the headworks landslide to protect intake operations. However, major risks, including those from the landslide, will remain.

Due to the extreme geomorphological and hydrological processes in the catchment, it would be impossible to control all the landslides, floods, and debris flows that threaten the current location of the headworks. For the long term, therefore, the relocation of the headworks to an area with lower risk from devastating hazards should be considered. Due to the upper catchment risks and uncertainties, the following approach for designing the headworks is proposed:

- A **phased and observational approach** to manage the inherent uncertainties in hydraulic design criteria, and by which components can be added in a stepwise manner based on the performance observation.

- A **performance-based design** with flexible design criteria for individual components such that the most critical elements are robust and able to survive extreme events, while the less critical components can be repaired after the event. This approach optimizes the design and ensures cost-effectiveness.

- A design principle to **enhance system redundancy**.

River Basin Scale

The events of 2021 were the result of interacting hazards including heavy monsoon rains, landslides, flash floods, debris flows, and debris floods. Climate change further aggravated glacier dynamics and caused snow cover and permafrost to play a role in the cascading string of events.

Due to the complexity of these hazards, installation of an EWS is essential. Besides protecting the water supply infrastructure, this system would cut loss of life and reduce the impact on the livelihoods of people living alongside the Melamchi River.

Given the wide-ranging and costly impacts of the 2021 floods, the response to such disasters cannot, and should not, be borne by the water supply sector alone. Agriculture, agroforestry, hydropower schemes, businesses, and homes were affected, while the destruction of roads and bridges cut off villages and hampered the transport of food and supplies. Therefore, an IRBM strategy, which involves all affected sectors to collectively investigate, plan, and develop strategies for best use of land, water, and other resources under improved hot spoteconomic and environmental conditions, should be a priority.

Regional Scale

Other Himalayan countries, including Bhutan, India, and Pakistan, are experiencing similar levels of flooding hazards that are often triggered by common factors, have the same types of impacts, and require the same analyses and mitigation measures. Many areas of these countries are also prone to earthquakes and face increasingly intense weather events as climate change affects the permanent snowline, extends monsoon rains, and causes heat waves.

As national agencies struggle to improve monitoring, step up DRM, and understand the impacts of climate change, regional initiatives could help better pool data, best practices, and set up knowledge repositories. ADB has already taken the first steps to improve regional coordination through technical assistant support which could include sharing research, EWS, catchment climate risk assessment guidelines, and design criteria. The guidelines, norms, and standards that will emerge from these initiatives will provide a broad and well-founded platform for developing new projects in the region.

REFERENCES

Arnold, J. et al. 1998. Large-area hydrologic modeling and assessment part I: Model development. *Journal of American Water Resources Association*, vol. 34/1.

Ashrit, R. 2010. *Investigating the Leh 'Cloudburst'*. https://www.ncmrwf.gov.in/Cloudburst_Investigation_Report.pdf

Asian Development Bank (ADB) and Melamchi Water Supply Development Board (MWSDB). 2022. *Damage assessment of Melamchi Headworks and Hazard Mapping of the Catchment*, TA6596-NEP, Inception Report.

Bajracharya, A., Bajracharya, S., and Shrestha, A. 2018. Climate Change Impact Assessment on the Hydrological Regime of the Kaligandaki Basin, Nepal. https://www.sciencedirect.com/science/article/pii/S0048969717337658

Bense, V.F., Kooi, H., Ferguson, G. and Read, T. 2012. Permafrost Degradation as a Control on Hydrogeological Regime Shifts in a Warming Climate. *Journal of Geophysical Research: Earth Surface*, 117(F3).

Connon, R. et al. 2014. Changing hydrologic connectivity due to permafrost thaw in the lower Liard River valley, NWT, Canada. *Hydrological Processes*, 28 (14): 4163-4178.

Department of Mines and Geology, NDRRMA. 2021. *Melamchi Disaster Preliminary Field Investigation*. https://www.researchgate.net/publication/355166405_Melamchi_Disaster_Preliminary_field_investigation_Report_DMG_and_NDRRMA_2021

Dhital M.R., 2015, *Geology of the Nepal Himalaya*. Springer International Publishing, Switzerland, 498 p.

Dhital M.R., Sunuwar S.C., and Shrestha, R 2002. Geology and structure of the Sundarijal–Melamchi area, central Nepal. *J Nepal Geol Soc 27 (Special Issue)*–10.

Dijkshoorn, J.A., Huting, J.R.M., 2009. *Soil and Terrain (SOTER) Database for Nepal. Report 2009/01.* Accessed 15 December 2016 (Online Dataset).

Explorer Geophysical Consultants Pvt. Ltd. 2022. *Report on Geophysical Investigation using Electrical Resistivity Tomography Survey (2D-ERT) and Seismic Refraction Tomography Survey (2D-SRT) at the Bhemathan Deposit*. Subcontractor to TA 6596-NEP for ADB and MWSDB.

Gautam, D. and Rupakhety, R. 2021. Empirical Seismic Vulnerability Analysis of Infrastructure Systems in Nepal. *Bulletin of Earthquake Engineering*. 19. 10.1007/s10518-021-01219-5.

Hagen, T. 1969. *Report on the Geological Survey of Nepal. Volume 1. Preliminary Reconnaissance.* OCLC Number/Unique Identifier: 898846648

HEC-RAS, River Analysis System. *HEC-RAS River Analysis System User's Manual Version 6.0. February 2016.* https://www.hec.usace.army.mil/software/hec-ras/documentation.aspx

ICIMOD. 2016. *The Impact of Nepal's 2015 Gorkha Earthquake-Induced Geohazards.* ICIMOD Research Report 2016/1. ICIMOD, Kathmandu, Nepal.

IUCN. 1999. *Melamchi Diversion Scheme Environmental Impacts Assessment (EIA) Report.* International Union for Nature Conservation (IUCN), Kathmandu, Nepal.

Jianchu, X. et al. 2007. *The melting Himalayas: Regional challenges and local impacts of climate change on mountain ecosystems and livelihoods.* International Center for Integrated Mountain Development (ICIMOD).

Mishra, V., Bhatia, U., and Tiwari, A. 2020. *Bias-corrected climate projections from Coupled Model Intercomparison Project-6 (CMIP6) for South Asia.* http://arxiv.org/abs/2006.12976

NDRRMA/World Bank Study. 2021. *Final Report on Aerial Image Acquisition Using Drone Flights of Melamchi, Helambu and Panchpokhari Area (Indrawati and Melamchi River) of Sindhupalchowk.* By consultant, Trimax IT Infrastructure & Services Pvt Ltd.

NDRRMA/World Bank Study. 2021. *Melamchi Flood Modeling.* Final report.

Neitsch, S. L. et al. 2002. *Soil and Water Assessment Tool: The Theoretical Documentation, Version 2000.* Texas Water Resources Institute, College Station, Texas, TWRI Report TR-191.

Pamungkas, B., Misra, P., Kambhampati, P., Takami, J., Vicente, F., Chavhan, R., and Arya, A. 2021. *InSAR-based Slope Susceptibility Characterization in Melamchi River Upstream Area.* Synspective Inc.

Roback, K. et al. U.S. Geological Survey. 2017. https://doi.org/10.5066/F7DZ06F9

Shrestha, A. 1998. *Estimation of Evapotranspiration by Penman Monteith Equation and Analysis of General and Extreme Rainfall over Kulekhani River Catchment.* Tribhuvan University, Kathmandu, Nepal.

USACE. 2015. *Hydrologic Modeling System (HEC-HMS User's Manual).* Hydrologic Engineering Centre (HEC), US Army Corps of Engineers (USACE).

WMO. 2009. *Manual on Estimation of Probable Maximum Precipitation (PMP).* https://library.wmo.int/index.php?lvl=notice_display&id=1302

Yao, T., et al. 2020. *Third Pole Climate Warning and Cryosphere System Changes.* https://public.wmo.int/en/resources/bulletin/third-pole-climate-warming-and-cryosphere-system-changes

www.ingramcontent.com/pod-product-compliance
Lightning Source LLC
Chambersburg PA
CBHW050046220326

41599CB00045B/7306